Astronomers' Universe

Other titles in this series

Origins: How the Planets, Stars, Galaxies,
and the Universe Began
Steve Eales

Calibrating the Cosmos: How Cosmology Explains Our
Big Bang Universe
Frank Levin

The Future of the Universe
A. J. Meadows

It's Only Rocket Science: An Introduction to Space
Enthusiasts (forthcoming)
Lucy Rogers

Rejuvenating the Sun and Avoiding Other
Global Catastrophes
Martin Beech

Maurizio Gasperini

The Universe Before the Big Bang

Cosmology and String Theory

 Springer

Prof. Maurizio Gasperini
Università di Bari
Dipartimento di Fisica
Via G. Amendola, 173
70126 Bari
Italy
gasperini@ba.infn.it

ISBN: 978-3-540-74419-1   e-ISBN: 978-3-540-74421-4

Library of Congress Control Number: 2008927472

Cover design: deblik, Berlin

Printed on acid-free paper

9 8 7 6 5 4 3 2 1

springer.com

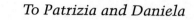

To Patrizia and Daniela

Preface

The idea of preparing this book grew out of a series of lectures and seminars held over several years in various Italian universities. The interest aroused in the students – and in colleagues not specialized in the field, who were also present at the talks – led me to the idea of writing a non-technical introduction to the newly-born field of string cosmology, aimed at a wider range of readers than just the professional community who usually attend the international conferences and read the specialized journals.

The challenge with this book is to present new possible scenarios for the primordial Universe emerging from recent developments in theoretical physics, but without resorting to too many numbers and equations, and using instead a series of illustrative cartoons. The book is addressed, in particular, to all those readers with at least a basic (high-school) knowledge of physics, but not necessarily equipped with an academic scientific background.

As a consequence, the discussion of many issues will be qualitative, often incomplete, and sometimes even grossly approximate. Nevertheless, I hope that the introductory picture provided by this book will be detailed enough to enable the reader to understand the most recent cosmological models, the key underlying ideas and, above all, how they can be tested using the experimental tools provided by current technology.

The physical grounds for such ideas are deeply rooted in the so-called theory of strings (or *string theory*, for short). Within modern physics, string theory provides in principle a robust theoretical framework for a complete and unified description of all the forces of Nature, at all energies – actually, it is at present the *only* theoretical scheme able to include the gravitational force in a consistent way, even in the quantum regime. One of the possible consequences of string theory is a cosmological scenario in which the great initial deflagration commonly called the *Big Bang* may not necessarily coincide with the birth of our Universe; rather, it could represent just an intermediate step in the whole history of the cosmos. Given the potential relevance of this picture (and the possible impact even beyond its strictly scientific applications), it is probably appropriate to attempt to put it across to a non-specialist audience.

The present version of the book is partly based on an earlier Italian edition, which has been extensively brought up to date taking into account the most recent – theoretical and experimental – developments in the physics of the early Universe. I should mention, in particular, the latest (2006) results of the WMAP satellite on the experimental side, and the inflationary scenarios based on brane interactions on the theoretical side. In addition, the former edition has been completed by new figures and new important explanatory parts concerning string theory and its revolutionary impact on our understanding of fundamental physics.

It is a pleasure, as well as a duty, to thank the many researchers with whom I have worked over the years on various aspects of string cosmology, and whose collaboration I hope to continue. They are, in alphabetical order: Luca Amendola (Observatory of Rome, Italy), Valerio Bozza (University of Salerno, Italy), Ram Brustein (Beer Sheva University, Israel), Alessandra Buonanno (University of Maryland, USA), Cyril Cartier (University of Geneva, Svitzerland), Marco Cavaglià (University of Mississippi, USA), Eugenio Coccia (University of Rome "Tor Vergata", Italy, currently Director of the Gran Sasso National Laboratory, L'Aquila, Italy), Edmund Copeland (University of Nottingham, UK), Giuseppe De Risi (University of Bari, currently at the University of Portsmouth, UK), Ruth Durrer (University of Geneva, Switzerland), Massimo Giovannini (University of Turin, Italy, currently at CERN, Switzerland), Michele Maggiore (University of Geneva, Switzerland), Jnan Maharana (Bubaneshwar University, India), Kris Meissner (University of Warsaw, Poland), Slava Mukhanov (University of Munich, Germany), Stefano Nicotri (University of Bari, Italy), Federico Piazza (University of Milan "Bicocca", Italy, currently at the Perimeter Institute for Theoretical Physics, Canada), Roberto Ricci (University of Rome "Tor Vergata", Italy), Mairi Sakellariadou (University of Athens, Greece, currently at King's College, London, UK), Norma Sanchez (Observatory of Paris, France), Domenico Tocchini-Valentini (Observatory of Rome, Italy, current at The Johns Hopkins University, Baltimore, USA), Carlo Ungarelli (University of Pisa, Italy), and Gabriele Veneziano (Collège de France, Paris). Beside these people there are many other scientists who have originally and independently contributed to the cosmological models presented in this book, and to whom I will make reference in the subsequent chapters (see also the website dedicated to string cosmology available at the address http://www.ba.infn.it/~gasperin).

I would also like to thank the various national and international scientific collaborations that have kindly permitted the use of figures and photos regarding gravitational wave and cosmic microwave experiments. I am grateful, in particular, to the following scientists (in alphabetical order): Peter Bender (University of Colorado, USA, on behalf of the LISA collaboration), Massimo Cerdonio (University of Padua, Italy, on behalf of the AURIGA collaboration), Adalberto Giazotto (INFN Pisa, Italy, on behalf of the VIRGO collaboration), and Jan Tauber (ESA Astrophysics Division, on behalf of the PLANCK collaboration).

However, there are not enough words for thanking my collaborator and friend Gabriele Veneziano, former staff member (and former Director of the Theory Division) of the European Center for Nuclear Research (CERN) in Geneva, Switzerland, now Professor at the Collège de France, in Paris. Gabriele started the original project for this book with me, but unfortunately was unable to pursue it due to later commitments. Despite that, he has generously helped me to write the chapter specifically devoted to strings – and indeed, he is a world-renowned expert on strings, besides being one of the founding fathers of string theory – and his advice has also been invaluable in many other parts of the book. It is fair to say that this book would not exist in its present form without his original contributions and the passionate commitment to research that we have shared over many years. So any credit for the book is also partly his due, while I assume full responsibility for any imperfections.

Last but not least, I am very grateful to Angela Lahee (Physics Editor at Springer) for her kind encouragement and advice, and for many important suggestions. I am also grateful to Carlo Ungarelli for his careful translation of the original Italian manuscript. Finally, special thanks are due to my wife Patrizia and my daughter Daniela. Besides their continuous support and encouragement they also helped me, as potential target readers, providing useful suggestions on how to improve in many points the first draft of the manuscript.

Cesena,
December 2007   *Maurizio Gasperini*

Contents

1. Introduction

The past century has been characterized by ever-increasing progress in our knowledge of nature and our understanding of its physical laws. The experimental investigation of the properties of matter, starting from the development of atomic physics at the end of the nineteenth century, has allowed us to look inside the atom, inside its nucleus, and even inside the constituent particles of the nucleus, pushing the frontier towards ever-decreasing distances and ever-increasing energies. At the opposite scale, astronomical and astrophysical observations have allowed us to go beyond the frontiers of our solar system and our galaxy, and we have even broken free from every kind of optically active system, pushing the frontier towards ever-increasing distance scales and thereby exploring older and older epochs.

At the same time, the development of progressively more sophisticated theoretical and mathematical models such as relativity, quantum mechanics, and field theory, has allowed us to build up a coherent framework to accommodate and understand this vast amount of experimental data. The two paths laid down by the development of nuclear physics and astrophysics, apparently divergent (in distance scale) but effectively convergent towards ever-increasing energies, then successfully merged, yielding, during the 1970s, the so-called standard cosmological model. It is certainly not an overstatement to say that this model represents one of the pillars of twentieth-century physics.

The standard cosmological model, which will be described in detail in the following chapters, provides us with a complete and satisfactory description of the current Universe. Furthermore, this model can be extrapolated backward in time to recover the temporal evolution of the Universe – explaining for instance the origin of light elements (so-called nucleosynthesis), starting from an initial state characterized by a primordial hot "mixture" of elementary particles. Moreover, the natural completion of the standard model, known as the inflationary model, explains how the large scale structures that we currently observe (galaxies, clusters of galaxies) may emerge from tiny primordial fluctuations in the matter density.

According to the standard model and its "inflationary" extensions, the Universe is a system which has continously expanded from a huge initial explosion, commonly known as the Big Bang. The relics of this explosion (in particular, the cosmic microwave background, electromagnetic radiation characterized by a thermal, black-body spectrum) was first observed in 1965 by Arno Penzias and Robert Wilson, who were awarded the Nobel Prize for this discovery. Despite the fact that such results are relatively recent, the concept of the expanding Universe has already become part of popular culture. Indeed, expressions like "explosive Universe", "Big Bang", and "initial singularity" are now common language. There is widespread awareness that the Universe is "expanding". A number of excellent popular science books, written by world-renowned scientists, describe the history of the Universe from the Big Bang to the present time.[1]

But what exactly do we mean by the Big Bang?

As the term suggests, a Big Bang is certainly a big explosion. More precisely, a rather violent and fast production of radiation and matter particles characterized by extremely high density and temperature. The cooling produced by the expansion (according to the standard laws of thermodynamics) has "firmed up" such particles into matter lumps, that have eventually combined into the large scale structures of the Universe we observe today. We can say that these aspects of cosmological evolution are well understood and widely accepted, barring some still debated issues concerning, for instance, the problem of baryogenesis (i.e., the mechanism by which only matter particles are produced from the relics of the primordial explosion, while large lumps of antimatter seem to be completely absent today on large scales).

The term "Big Bang", however, is often used (even in a scientific context) in a broader sense, as synonymous with the birth and origin of the Universe as a whole. In other words, this term is used

---

[1] See for instance S. Weinberg: *The First Three Minutes* (Basic Books, New York 1977).

also to indicate the single event from which everything (including space and time themselves) directly originated, emerging from an initial singular state, i.e., a state characterized by infinitely high values of energy, density and temperature.

This second interpretation is certainly suggestive, and even scientifically motivated within the standard cosmological model. Nonetheless, it has been challenged by recent developments in theoretical physics that took place at the end of the twentieth century.

Indeed, recent theoretical progress[2] suggest that the behavior of matter at very high energies could be radically different from what we usually observe in the ordinary macroscopic world. In particular, when the energy and the corresponding strength of the various forces are very close to a critical value – to be defined later in the book – it may no longer be legitimate to describe matter in terms of point-like particles (as suggested by the well established laws of low-energy physics). Matter could in fact take more "exotic" forms, either thread-like (called strings) or membrane-like, thus occupying spatial patches that progressively increase with energy. Furthermore, an even more astonishing consequence of this scenario – to be discussed in Chap. 10 – is that, as the energy and strength of the forces increase, the effective number of dimensions of space also rises. In other words, the dimensionality of space-time is not rigidly fixed, but becomes a *dynamical variable*.

These new theoretical ideas therefore suggest novel descriptions of the initial state of the Universe. Close to the Big Bang, i.e., in a regime of very high energy concentration, the state of the Universe was quite different not only from its current state, but probably also from the one predicted by the standard cosmological model. Besides being extremely hot and dense, and highly curved, the Universe was probably also a higher-dimensional structure, inhabited by exotic objects like strings and membranes, and dynamically governed by forces and symmetry laws that have left today only extremely weak (and possibly indirect) traces.

Within this scenario, more flexible and richer than the standard one, it becomes possible to build cosmological models without

---

[2] See B. Green: *The Elegant Universe* (Vintage, London 1999) for a popular introduction.

any initial singularity, where cosmological evolution can be traced arbitrarily far back in time, even to infinity. Such models allow the Universe to exist, and develop through a long "prehistory", even before the actual Big Bang, now identified as the explosion which gives rise to the matter and to the forms of energy that we now observe. The Big Bang is still present but, although it remains a milestone in the evolution of the cosmos, no longer represents the origin of space, time, and the Universe itself. It thus becomes possible, within this framework, to explain *how* the Big Bang takes place, by studying mechanisms able to concentrate enough energy in a given space-time point to trigger the observed explosion.

All these aspects of modern cosmological models will be presented and illustrated – albeit in an incomplete fashion, if only due to lack of space – in the following chapters. But let us start by explaining how the hypothesis that the Big Bang was the origin of "everything", while having solid scientific roots, can nevertheless be challenged by recent developments in theoretical physics.

To this end, it is worth recalling one of the greatest lessons that the natural sciences have learned from Galileo, Newton, and the other founding fathers of modern physics: celestial bodies do not have any "mystic" essence or "metaphysical" property, but move and evolve in time according to the same laws that govern the dynamics of more mundane material objects. The whole Universe is itself an ordinary physical system obeying those laws that science seeks to discover and to piece together using reproducible experiments. The Universe that we observe today, in particular, can be fully (and satisfactorily) described on large scales by the laws of classical physics, including general relativity, the relativistic theory of gravitation developed by Albert Einstein at the beginning of the twentieth century. This theory both includes and generalizes Newton's gravitational theory, and has successfully passed all experimental tests performed since its conception.

As will be discussed in the next chapter, the theory of general relativity predicts a warping of space and time which is directly proportional to the energy density distributed in the matter sources. By applying this theory to our expanding Universe one then obtains a cosmological model in which the curvature of the Universe itself evolves with time, following the corresponding evolution of the energy density and temperature.

As the expansion proceeds, matter becomes progressively more rarefied and colder, according to standard thermodynamics. Thus, as a consequence of general relativity, the curvature of the Universe becomes gradually smaller. It is intuitively obvious, in particular, that an infinite expansion would tend to render the Universe completely empty, and its geometry – i.e., the space-time, to use relativistic jargon – would tend to become flat. In a similar fashion, one can use general relativity to establish that, in the past, when the Universe was smaller and more compact, it was also hotter, denser and thus much more warped than it is today. Going progressively backward in time the density, the temperature, and the curvature of the Universe increase without bound until they reach – in a long, but finite time interval – an infinitely dense, hot and curved "singular" state.

The idea that such a singular state (identified with the Big Bang, and conventionally placed at the time coordinate $t = 0$) may represent the birth of the Universe is based upon the fact that the dynamical equations of general relativity lose their validity at the onset of a singularity, and cannot be extended beyond a singular point (in this case, backward in time beyond $t = 0$). In other words, the solutions of those dynamical equations describe an "incomplete" space-time which is not infinitely extended in time, being characterized by an impassable "boundary" located at a finite temporal distance from any physical observer. It is thus general relativity itself which, in a cosmological scenario, unavoidably leads to the notion of an initial singularity, enforcing the idea that the Big Bang was the beginning of space-time and the moment of birth of our Universe.

If we were to adhere strictly to general relativistic predictions, we should then conclude that the main topic of this book – the Universe before the Big Bang – is something meaningless. Before jumping to this conclusion, however, there is a question we should ask ourselves. Is the incompleteness of space-time predicted by general relativity a true physical property of our Universe, or is it only a mathematical property of some equations, that are really inadequate to describe space and time near the Big Bang?

This is certainly a legitimate question in physics, where the occurrence of a singularity often does not correspond to any real entity, but is just a signal that some physical laws have been extrapolated beyond their realm of validity.

Let us consider the following, very simple and well-known example. The laws of classical electromagnetic theory establish that, inside the atom, the positively charged nucleus exerts an attractive force on the negatively charged electron, and that this mutual force increases as the distance between the two charged particles decreases (according to the well-known Coulomb law). In particular, when the distance between the nucleus and the electron tends to zero, the force becomes infinite. On the other hand, a revolving electron should progressively radiate away its energy, thus progressively shrinking its orbit closer and closer to the nucleus. We should then conclude that, according to the classical electromagnetic laws, all electrons would eventually fall into the nucleus, atoms would collapse into singular point-like states, and ordinary matter would not exist in the form we know it. Such a situation does not occur, however, simply because at short enough distances the laws of classical physics break down and the laws of quantum mechanics come into play, preventing the collapse of the electron into the nucleus.

We may also refer to another example, less obvious, but equally well known to physicists. The energy density of thermal radiation, computed by applying the laws of classical physics, obeys the so-called Rayleigh–Jeans spectrum. This predicts an unbounded growth of the energy density with the frequency of the thermal radiation. But once again this energy singularity disappears if we take into account the need to use quantum mechanics to describe the behavior of matter and radiation at high enough frequencies (i.e., at high enough energies). One then finds, by applying the required quantum mechanical principles, that the thermal energy density first increases with frequency, reaches a maximum at a finite frequency value, and eventually decreases as the frequency goes to infinity, following the so-called Planck spectrum (named after Max Planck, who was one of the founders of quantum mechanics).

There are also other circumstances, however, where the occurrence of a singularity in the equations describing a physical system may point to some abrupt change in the state of the system, requiring the introduction of different variables and different degrees of freedom for an appropriate description. In this case it is also instructive to consider a simple example, drawn from particle physics.

Let us first recall that at sufficiently low energies (i.e., well below the typical energy of strings) all known ordinary matter – including also those forms of matter produced artificially in the various accelerators around the world – can be reproduced by a proper combination of a relatively small number of fundamental building blocks, the so-called elementary particles. Some of these particles (actually, only a very small fraction of their total number) are stable: this means that, were they set up in a fully isolated environment, they would persist in their original state, retaining their physical properties unchanged for an infinitely long time. Other particles, however, are unstable: even without any external influence, these particles decay, that is, they disappear, leaving in their place two or more different (and lighter) particles. Their mean decay time, called the lifetime, depends upon the forces producing this intrinsic instability.

Consider, for instance, an atom. It consists of electrons (which are stable particles), protons (which are also stable, as far as we know) and neutrons, which are stable, but only within the atomic nucleus. In an empty environment (i.e., in vacuum) a neutron decays, with a typical lifetime of the order of fifteen minutes, producing three new stable particles: a proton, an electron, and a neutrino. Now for each of these "newly born" particles, the decay process can be regarded as a kind of "Big Bang" in the realm of subnuclear physics: an abrupt explosive process marking the appearance of these particles and the beginning of their life, on a microscopic scale. This does not imply, however, that these particles emerged from "nothing". Before they appeared, there was a corresponding physical system in a different initial state, representing a neutron which, under the influence of some nuclear forces (called weak interactions and first described theoretically by Enrico Fermi), has transformed into a new state represented by three different particles.

There is no doubt that the physical description of the system undergoes a sudden and abrupt change when the neutron decay occurs. Nonetheless, the decay itself does not represent any impassable boundary. In a similar fashion, the cosmological explosion that we identify as the Big Bang certainly marks the beginning of the present state of the Universe, i.e., of the Universe in the form that we currently observe. However, if we relax the a priori assumption that the Big Bang must also mark the origin of space and time, the question as to whether our Universe existed before such an explosive event, and in which state, may become perfectly meaningful.

An equally legitimate question, however, could also naturally arise at this point: Why should we address the issue about a possible state of the Universe before the Big Bang, thus casting doubts on the hypothesis – suggested and supported by general relativity – that the Big Bang is effectively the true beginning of everything?

The answer to this is quite simple. General relativity, as previously stressed, is a classical theory. It has been successfully tested at densities, temperatures and curvatures much higher than those we may observe in our ordinary macroscopic world, but definitely much lower than the ones coming into play in the primordial Universe. The use of general relativity near the Big Bang implies trusting the validity of this theory not only beyond any experimental evidence, but also in a regime where there are well-founded reasons for doubting the legitimacy of classical theories.

In fact, in the regime of extremely high energies, where the above-mentioned strings and membranes may become relevant, the properties of the gravitational interaction are expected to be significantly different from those predicted by general relativity. New fields and new kinds of short-range interactions may come into play, as inevitable consequences of the laws of quantum physics. On the other hand – as we shall see in the following chapters – it is the standard cosmological model itself that leads us to the unavoidable conclusion that quantum mechanics, together with the physical laws appropriate to describe matter on microscopic scales, are key elements in the dynamics of the primordial Universe.

Taking the expansion of the Universe seriously, and going backward in time, we do indeed reach epochs during which the entire structure of the Universe and its energy (currently spread over billions of galaxies) was compressed into a spatial region of about one hundredth of a millimeter in length. The energy density of the Universe at that time was inconceivably high compared to what we usually observe on macroscopic scales. We can compute, using general relativity, that the energy density for such a small compact region was about 10^{80} times greater than the typical density in an atomic nucleus (which is already very high). Such a value, dubbed the Planckian limiting density, is the threshold value corresponding to the onset of a regime where the geometry of space and time itself (together with matter) must obey the laws of quantum mechanics. General relativity, however, does not know about quantum mechanics: it can thus bring us to the doorstep of the Big Bang, so to speak, but it cannot proceed further without entering a regime in which its predictions are no longer reliable.

Therefore, in order to correctly describe the Universe when approaching the Planckian regime, a classical theory like general relativity is not sufficient. Instead one must have a theory able to provide a consistent description of gravitation even within a quantum framework. Since such a theory was not available when the standard model was developed, speculative attempts were made to extrapolate the predictions of general relativity right to its limits, that is, to describe the birth of the Universe from an infinitely hot, dense, and curved state: the initial singularity, beyond which nothing existed.

This methodology wherein the results of a known theory are extrapolated into an as yet unexplored range is a natural procedure after all, and it is common practice in the scientific context, as a first step towards more sophisticated theories and more complete models. However, as far as cosmology is concerned, pushing this procedure to its extreme leads us to identify the limits of our current knowledge with a natural barrier, as though nature had set up a definitive, impassable gate at the Big Bang position. Such a situation is reminiscent of the attitude the ancient peoples had towards the Columns of Hercules: since no-one had crossed the strait of Gibraltar, and no-one knew the world beyond it, it was common opinion (and it seemed plausible) that the world would end at that point.

This basic lack of knowledge, however, is continuously being filled by the recent developments of theoretical physics, which have provided us with a very powerful tool: string theory. In principle, this theory (and its possible, though as yet not fully defined completion, M-theory) allows a coherent merger of quantum mechanics and gravitation, and thefore provides a potentially consistent framework to describe the geometry of space-time in the regime of extremely high energy densities and curvatures. It has thus become possible to study the evolution of the Universe near the Big Bang, and even beyond it, by means of a robust and consistent theory, valid at all energies. It is as though, in the above analogy with ancient times, someone had built a more solid and reliable ship that would allow some brave explorers to sail the seas beyond the Columns of Hercules. In this way, it has been found that the extension of space-time is not necessarily constrained by an initial singularity, and questions about the possible state of the Universe before the Big Bang are fully legitimate and well posed.

Anticipating the demand of the curious reader, and as an introduction to the content of the following chapters, let us immediately give some idea of what the Universe would look like according to the indications provided by string theory, if we could look back in time to the epoch of the Big Bang, and even beyond the Big Bang itself. Such remote epochs cannot be traced using objects like stars and galaxies, which formed only very recently on the time-scale of cosmic evolution. These structures were not yet formed at the onset of the Big Bang, and neither did they exist before it. Instead, we need to exploit some geometrical properties of the Universe that are always valid, like space-time curvature. Let us therefore ask about the past evolution of space-time curvature, and represent its behavior graphically as a function of time.

According to the so-called standard cosmological model (which will be introduced in Chap. 2, and which is the model providing the grounds for the hypothesis of the Big Bang as the singular beginning of "everything") the Universe expands and the curvature decreases in time in a continuous and decelerating fashion. Hence, going backward in time, we reach epochs characterized by progressively increasing curvature. This monotonic growth proceeds continually until the infinite curvature state is reached (corresponding to a singularity, and conventionally fixed at the initial time $t = 0$). Beyond that point, no classical description is possible (see Fig. 1.1).

However, as already pointed out, a singularity can often be interpreted in a scientific context as a signal that we are applying

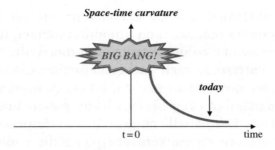

STANDARD COSMOLOGY

FIGURE 1.1 The *bold solid curve* describes the behavior of the curvature scale of our Universe as a function of time, according to the standard cosmological model. The further we go back in time, starting from the present epoch, the higher is the curvature, approaching infinity as t approaches zero. Thus $t = 0$ is identified with the moment of the Big Bang and the beginning of space-time itself

some physical laws outside their realm of validity. Concerning this point, it is interesting to quote the opinion of Alan Guth, one of the fathers of modern inflationary cosmology (a subject covered in Chap. 5). In his recent book[3] he makes the following remarks about the initial singularity:

> It is often said – in both popular-level books and in textbooks – that this singularity marks the beginning of time itself. Perhaps it's so, but any honest cosmologist would admit that our knowledge here is very shaky. The extrapolation to arbitrarily high temperatures takes us far beyond the physics that we understand, so there is no good reason to trust it. The true history of the universe, going back to "$t = 0$", remains a mystery that we are probably still far from unraveling.

In other words, according to Guth, there is little hope of describing the initial phase of the Universe within the standard cosmological model. Indeed, as we have already pointed out, in the presence of arbitrarily high curvature, energy and density, the Einstein theory of gravitation ceases to be valid, and the associated description of the space-time geometry becomes meaningless.

Beside the singularity problem, however, there are also other issues concerning the standard cosmological model that hint at the

---

[3] A. Guth: *The Inflationary Universe* (Vintage, London, 1997), p. 87.

need for a modification near the initial time, even before reaching the quantum gravity regime. Such a modification requires in particular that, at some point during its primordial evolution, the Universe should undergo a phase of highly rapid expansion, dubbed inflation. We are giving here just a glimpse of what will be illustrated in more detail in Chap. 5. For the purposes of our fast-track, time-reversed journey, it will be enough to point out that during an inflationary phase of conventional type the evolution of the Universe is expected to be determined by the energy density of a "strange" particle – dubbed the inflaton – that generates a scalar-type field strength.

Going further backward in time, the potential energy of this field progressively increases, and eventually becomes so strong as to be able to "freeze out" the space-time curvature. Then, as shown in Fig. 1.2, the curvature of the Universe stops increasing and levels off to an almost constant value. During this initial inflationary phase, the geometry of the Universe thus approaches that of the de Sitter space-time (named after the cosmologist who found the solution describing a spacetime with constant curvature). The primordial Universe, in that case, closely resembles a tiny, four-dimensional hypersphere with constant radius.

However, there is also a problem in this case: a phase in which the Universe expands while the curvature stays fixed at a constant value cannot be extended backward in time without limit. In fact,

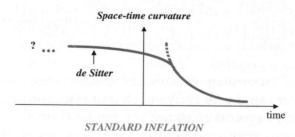

FIGURE 1.2 The *bold solid curve* describes the behavior of the curvature scale of our Universe as a function of time according to the conventional inflationary model. When the Universe enters the inflationary regime the space-time curvature, instead of growing as predicted by the standard cosmological model (*dashed curve*), tends to become frozen at a constant value, asymptotically approaching a phase associated with a de Sitter geometry

for a physical (stable) particle, moving according to the laws of general relativity within this type of geometry, it would take a long, but certainly finite amount of time to reach us starting from the moment when the radius of the Universe was zero. In order to obtain a "complete" model of space-time, the initial Universe should exist in a contracting phase, at least according to the de Sitter solution to the equations of general relativity. However, as has been shown by some cosmologists (in particular, Arvind Borde, and Alexander Vilenkin), within the framework of an inflationary model based upon the potential energy of some scalar field, a transition between a contracting and an expanding phase is not allowed, at least according to general relativity and the physical laws that we currently believe to be valid.

Hence, the inflationary scenario at constant curvature is also unable to provide a complete model for the evolution of our cosmos. As Guth himself points out in his book[4]:

> Nonetheless, since inflation appears to be eternal only into the future, but not the past, an important question remains open: How did it all start? Although eternal inflation pushes this question into the past, and well beyond the range of observational tests, the question does not disappear.

Apart from the possibility of experimental tests (which will be discussed later in the book) it seems undeniable, as outlined by Guth, that a constant curvature, expanding phase cannot be arbitrarily extended backward in time, and does not lead to a complete description of the origin of our Universe. A possible solution to this problem (which will be introduced and discussed in Chap. 8) relies upon the possibility that a Universe with the appropriate, expanding de Sitter geometry might spontaneously emerge from the vacuum at some very early epoch (but not infinitely distant in time), through a typical quantum mechanical effect.

Leaving aside this possibility for the moment and limiting our options to a classical context, it is clear that if the curvature cannot remain indefinitely constant, then we are only left with two alternatives in order to extend our cosmological description

---

[4] A. Guth, *op.cit.*, p. 271.

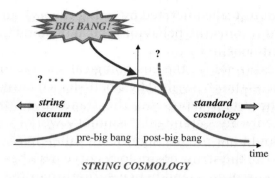

FIGURE 1.3 The *solid bell-shaped curve* describes the behavior of the curvature scale of our Universe as a function of time according to a typical
string cosmology model. The phase of maximal, finite curvature at the top
of the bell replaces the singularity of the standard scenario and describes
the Big Bang as a moment of transition between growing and decreasing
curvature. The curve interpolates between a pre-Big-Bang phase, describing the initial evolution from the vacuum state of string theory, and a
post-Big-Bang phase, evolving according to standard cosmological predictions

back in time. The first possibility is that at some point the curvature starts to increase again. In this case, however, the singularity would persist, with the only difference that the position of
the Big Bang would be moved backwards in time with respect to
the standard cosmological model. The second possibility is that,
going backwards in time, the curvature starts to decrease, becoming smaller and smaller as we go back in time. This second case is
exactly the model of the Universe suggested by string theory, represented in Fig. 1.3, which will be discussed in detail in the following
chapters.

In fact, as will be shown in Chap. 3, string theory suggests
that the plot of the cosmological curvature scale versus time could
have a specular reflection symmetry with respect to the time coordinate $t = 0$. It also suggests models in which there is no singularity in the space-time curvature, and time can be arbitrarily
extended to infinity, in both the backward and the forward directions. Within these models – dubbed string cosmology models, as
opposed to standard cosmology or conventional inflationary models – the curvature starts from arbitrarily small values, increases

up to a maximum value (dictated by string theory), and eventually decreases until it joins the behavior typical of standard cosmology and the current epoch.

The typical trend for the cosmological space-time curvature, in this context, is thus described by a bell-shaped curve, eventually joining the curve that represents the standard-model curvature. Indeed, at recent times standard cosmology works well, so string cosmology should not predict significant differences. However, as shown in Fig. 1.3, the string cosmology curve matches the standard one at higher curvatures (and hence *earlier*) than the curve corresponding to the inflationary de Sitter model. This occurs because the highest value of curvature reached in a string cosmology context (viz., the top of the bell curve drawn in Fig. 1.3) is in general higher than the value reached during de Sitter inflation. This feature may have some key phenomenological consequences, as will be discussed in the following chapters.

The moment at which the curvature reaches its (high, though finite) maximum value at the top of the bell replaces the singularity and corresponds to the position of the Big Bang in the standard cosmological scenario. It is then natural to refer to the cosmological phase characterized by increasing curvature (the left-hand side of the bell in the graph) as the pre-Big-Bang phase, describing the initial evolution of the Universe starting from an initial "vacuum" state (to be defined in Chap. 3). In the same way, the right-hand side of the bell corresponds to the post-Big-Bang phase, characterized by decreasing curvature and representing the typical evolution of the current Universe, in agreement with standard cosmological predictions.

According to this class of string cosmology models, the Universe at the epoch of the Big Bang was not a newly born baby, but a rather aged creature, midway through an evolution of probably infinite duration. Furthermore, the Big Bang itself is not viewed as a singular point, but as a transition – certainly violent and explosive, but of finite duration and intensity – between two phases characterized by different physical and geometrical properties. What happens, therefore, is that a traditional representation of our cosmos is somehow overturned, whence a comparison with the well-known Copernican revolution may seem natural. With Copernicus, the Earth lost its role as the center of the Universe, or focal point of the physical space. Similarly, within string cosmology, the Big Bang may lose its role as the beginning of the Universe, or focal point of physical time. A sort of Copernican revolution – in time, rather than in space – even though the Big Bang, in contrast to the Earth for Copernicus, does not completely lose its privileged role in the cosmic scene.

At this point, we are beginning to outline a potentially interesting cosmological scenario. However, before we proceed further, we cannot avoid asking the following question, which is of fundamental importance if we hope to keep working in a scientific context: What is the observational evidence that could either prove or disprove this scenario for the primordial evolution of the Universe?

The answer to this question is quite similar to the one an archaeologist would give to anyone asking him about the evidence supporting the existence of ancient civilizations. He would argue that, by studying old remains and available relics, one can attempt to trace back to original sources and reconstruct the past. In the same way, a cosmologist may be viewed as an archaeologist who studies the relics of the various cosmic epochs, in order to piece together the evolution of the Universe.

For a further clarification of this point let us go back to an example concerning unstable particles, and consider again the process of neutron decay introduced previously. In a sense, the particles left by the decay represent the relics of the decay process: analyzing such decay products, a physicist is able to trace back to and reconstruct the properties of the initial state. In particular, by studying the proton, the electron, and the neutrino that have emerged from the decay, and using his knowledge of the theory of weak interactions, he may deduce the prior existence of a neutron and compute, for instance, the neutron mass, even without having directly observed the initial particle.

In a similar fashion, the evolution of the Universe and the transitions between the various cosmological epochs are generally characterized by intense emission of a very large amount of radiation, of all kinds and in all allowed frequency bands. Part of such radiation has subsequently been transformed, by exploiting the relativistic equivalence between mass and other forms of energy. However, there is also a fraction of this radiation which reaches us today, retaining its original features. So by studying its properties, it is then possible to gain direct information about the past evolution of our Universe.

It is worth mentioning that, also in the context of the standard cosmological scenario, the hypothesis of a primordial explosion has been experimentally confirmed by the observation of relic electromagnetic radiation, the so-called cosmic black-body background, discovered by Penzias and Wilson. In the same way, as will be discussed in Chap. 6 and thereafter, other types of relic cosmic radiation – gravitational, dilatonic, or axionic backgrounds – could experimentally corroborate or rule out various scenarios describing the Universe before the Big Bang.

Finally, to complete this short introduction, let us just mention a few other topics that will be covered in subsequent chapters. A first and quite essential issue pertains to the motivations for introducing the cosmological phase already dubbed the pre-Big-Bang phase. Our aim, in particular, will be to explain how such a phase appears to emerge naturally within string theory, and not in the context of Einstein's gravitational theory.

Another point concerns the kinematical aspects of the primordial cosmological evolution. Despite their differences with respect to conventional inflationary models, string cosmology models are also initially characterized by a very fast, accelerated expansion. It follows that they are inflationary models too, in every respect, and may thus provide solutions for the shortcomings of the standard cosmological scenario.

But the point which is likely to be the key issue concerns the phenomenological consequences of these models, and the problems concerning their observation and their possible use as tests of string theory itself. We will therefore attempt to present a detailed description of the physical effects marking the differences between string models and more conventional cosmological models, paying particular attention to the possibility of either direct or indirect experimental observations of such differences.

To this end, it is interesting to observe that the phenomenological consequences of a cosmic phase preceding the Big Bang can be divided into three classes: type I, II or III (see Chap. 7). Type I includes observations that will be carried out in the near future (20–30 years from now); type II consists of observations relative to the immediate future (some years from now); and type III refers to observations (whether performed or not) that are already accessible to current technology. It was the very existence of experimentally testable consequences, emerging with growing evidence from the development of string cosmology models, that has encouraged and motivated many researchers to pursue the study of string cosmology and its various possibilities.

In order to provide the reader with a better introduction to the problems of cosmic evolution, and to the primordial epoch close to the birth of the Universe, it seems adequate to start with a concise overview of the standard cosmological model and its underlying theoretical background (the theory of general relativity). The next chapter will be devoted to this purpose.

2. General Relativity and Standard Cosmology

The physical description of the current Universe, and of the forces that stars, galaxies, and clusters of galaxies exert upon one another at the cosmic distance scales corresponding to their huge spatial separations, requires the use of general relativity, i.e., Einstein's gravitational theory. Some may ask why we do not just use Newton's gravitational theory. Why would it not be possible, for our Universe, to build up a consistent, Newtonian-type cosmology based upon the universal law of gravitation that we learnt at school, according to which the mutual attraction between two bodies is directly proportional to the product of their masses and inversely proportional to the square of their relative distance?

The answer to this question is very simple. It is based on the fact that Newton's theory is not a relativistic theory, and is thus valid only for sufficiently low velocities and energies. Strictly speaking, Newton's formulation is valid as long as the kinetic and potential energies of the bodies under consideration are small compared to the energy associated with their rest mass. This implies, in particular, that in order to apply the Newtonian theory correctly to a given system the associated potential energy per unit mass (the so-called gravitational potential) has to be much smaller than the square of the speed of light.

This requirement is certainly fulfilled by the gravitational forces that we ordinarily experience. This condition is satisfied, for instance, by the force exerted by our planet (the Earth) on us, and on its satellite (the Moon); it is satisfied by the forces by which the Sun holds onto its planets; it is even valid for the mutual forces holding together stars and galaxies. However, the above condition is *not* satisfied if we take into account the whole Universe accessible to our observation.

In fact, if we compute for the Universe the quantity that could represent the equivalent of the gravitational potential – multiplying

Newton's constant G by the total mass of the Universe (i.e., by the sum of the effective masses associated with all its cosmic components), and dividing by the radius of the portion of space containing these masses – the result we obtain is of the order of the square of the speed of light itself. Hence the need to resort to a fully relativistic theory of gravitation in order to formulate cosmological models able to consistently describe the dynamics of the Universe as a whole.

In the past, various attempts have been made to generalize Newton's theory in order to turn it into a relativistic theory of gravity. For instance, starting from the formal analogy between the Newtonian gravitational force (amongst masses) and the Coulomb force (amongst electric charges), attempts were made to describe gravity in terms of a four-dimensional vector field similar to the vector potential appearing in the relativistic theory of electromagnetism. These attempts failed for a fundamental reason: as is well known, electric charges of the same sign are mutually repulsive, while electric charges of opposite sign are mutually attractive (the same phenomenon also occurs with the poles of a magnet). In the gravitational case, however, there are no negative masses (to the best of our present knowledge), so that all masses have the same sign; nevertheless, we all know that masses attract each other, while for a vector-type gravitational theory they should repel each other, as happens in the case of electromagnetism.

Another approach attempted to describe gravity by using a relativistic scalar field to represent its potential, under the assumption that the gravitational potential energy retains its Newtonian form even when the typical speed of a body becomes relativistic. This attempt was also unsuccessful, for a number of reasons. In particular, the type of motion it predicts for the planets is inconsistent with observations. For instance, according to this scalar theory of gravity, the secular drift of the point closest to the Sun in Mercury's orbit (called the perihelion) would be much smaller – about one sixth of what is actually observed.

The correct path was actually taken by Einstein, nearly one century ago, when he came up with the idea of describing gravity in terms of a geometrical tensor field – a radically different approach to those used to describe the other known forces – and of formulating a consistent theory of gravity by extending and generalizing the fundamental principles underlying special relativity. Actually, this is the reason why his theory is called general relativity. In this theory, the well-known postulate of special relativity asserting that the physical laws are the same *in all inertial frames* generalizes into: the physical laws are the same *in all reference frames*, regardless of the coordinate transformations connecting them.

In other words, the physical equivalence of all inertial observers (i.e., those moving at constant speed) is extended to all observers, even to those whose motion is accelerated. This leads to what is known as the principle of general covariance, according to which the form of physical laws has to be invariant under any coordinate transformation – not only under the Lorentz transformations that connect inertial observers, and that represent the basic symmetries of special relativity. Strictly speaking, the principle of general covariance demands that the physical laws must be expressed as equalities between identical-rank tensors and that the latter, representing physical quantities like the energy, the force and so on, must be mathematical objects consistently defined with respect to generic coordinate transformations.

From a geometrical point of view, general covariance leads to a full-scale revolution in the geometrical structure of space-time: from the rigid structure of special relativity, which is of Euclidean type, we change to a deformable, generally curved structure, of Riemannian type (named after the mathematician Georg Riemann who first studied the geometry of such spaces). To visualize the differences, a table-top may be thought of as an example of rigid, two-dimensional Euclidean space, whereas an elastic net – which is flat when empty, but bends when massive objects are placed upon it – is an example of a two-dimensional (possibly curved) Riemannian space.

But why should general covariance, i.e., the physical equivalence between accelerated observers, require non-Euclidean geometry? A precise answer to this question would involve technical details that would take us beyond the scope of this book. To give an intuitive answer, it suffices to observe that in a Euclidean space the square of the distance between two points is given by the sum of the squares of the distances along the various axes of a Cartesian frame (as follows from a simple application of the well known theorem named after Pythagoras). If this is true in an inertial frame, then it remains true after performing coordinate transformations leading to any other inertial frame. However, this result is no longer valid, in general, when a coordinate transformation brings us to an accelerated reference frame. Here, to obtain the square of the distance between the two points, one needs to take the squares of the distances along the axes and, before making their sum, multiply them by non-trivial functions of the new coordinates which depend on the given transformation, and which represent the components of the so-called space-time metric.

Thus, an accelerated observer measures a space-time geometry which is not of Euclidean type. Instead, it is a Riemannian geometry, characterized by a metric which looks different in different reference frames. This explains why, extending the class of physically equivalent observers to include accelerations, one should expect a generalization of the geometry, and one should be ready to accept the possibility that the space-time is not generally flat. But what do the Riemann metric, or the curvature, have to do with gravity?

It is precisely the link between the space-time curvature and the gravitational interaction that is likely to represent the most novel feature of the theory of general relativity. In fact, in a curved space all test bodies tend to follow curved trajectories. Thus, their motion deviates from a straight line, *as if they were subject to forces*. Choosing a suitable metric (i.e., an appropriate space-time geometry), it then becomes possible to reproduce the gravitational forces, even in the Newtonian limit where the velocities are small.

In this way, one can directly incorporate the gravitational interaction into the space-time geometry. The latter is said to be flat (or Euclidean) in the absence of gravitational forces, while it is said to be curved when such forces are present. The equations of general relativity express this link between the space-time curvature and the gravitational properties of material bodies in a detailed fashion.

It is important to stress that this geometrization of forces works well in the gravitational case by virtue of the universality of gravity, i.e., the fact that all bodies "feel" gravity (and react to it) with the same intensity. Such universality is experimentally guaranteed by the well-known equality between the inertial mass and the gravitational mass. In particular, it is just this universality that underlies the so-called equivalence principle, according to which the effects of a gravitational field are *locally* indistinguishable at a given space-time point from those produced by a properly chosen accelerated frame. As a consequence, it is always possible to eliminate the gravitational field at a given point by means of an equal and opposite acceleration.

For other interactions, characterized by non-universal coupling constants, the geometrization of the corresponding forces would not be so effective and useful. Consider for instance the electromagnetic force. Since different bodies may have different electric charges, a geometric description would require associating different space-time geometries with the same electric field, depending on the test body upon which the force is exerted.

The geometrization of the gravitational field not only provides a new and elegant formalism for describing universal forces, but also has deep physical consequences and, in particular, predicts new gravitational effects (even in the low velocity regime) that are not present in the Newtonian theory of gravity. For the applications of this book, we particularly stress the slow-down (or time dilation) experienced by clocks in the presence of a gravitational field.

As already known within the theory of special relativity, the relative flow of time is different for observers who are not moving with the same velocity. A moving clock, in particular, is seen to tick more slowly if compared with an identical clock at rest. In a curved space, beside the slowing down due to the relative velocity, we may also measure a relative slowing down between two clocks at rest, provided they are placed at *different space-time positions*.

This phenomenon occurs because the Riemann metric which defines the distance between two space-time points, taking into account the possible curvature, deforms (with respect to the Euclidean case) not only the spatial intervals but also the temporal intervals. Now, suppose that between a given pair of points there is a difference in the gravitational field, and thus in the metric which describes the associated geometry. It follows that the relative distortion of the Euclidean time interval is also different. The comparison between these different intervals then leads to a relative slowing down of the clocks between the two different space-time points. In particular, one finds that the more the ticking of the clock is slowed down with respect to a clock in a flat space, the more warped the space is, i.e., the more intense is the gravitational field to which the clock is subjected.

For a periodic signal (e.g., an electromagnetic wave) propagating through a curved space, the time dilatation effect will thus produce an increase (with respect to the flat space) in the measured period, with a corresponding stretching of the wavelength (which is proportional to the period) and a decrease in the frequency (which is inversely proportional to the period).

This effect is commonly known as redshift – with reference to the spectrum of visible light, where a shift towards lower frequencies corresponds to a shift toward the red end of the color spectrum we know from the rainbow. This redshift is a peculiar example of a physical effect associated with the geometrical description of the gravitational field. Its experimental validation, as well as tests of other effects – such as the deflection and slowing down of light and electromagnetic signals in the presence of gravity, the correct prediction of the shift in Mercury's perihelion, etc. – have marked the success of general relativity as a consistent and phenomenologically viable relativistic theory of gravitation, which both completes and enriches Newton's theory. The gravitational redshift, in particular, leads to important applications in a cosmological context, as will be illustrated shortly.

In fact, assuming that general relativity is valid on length scales corresponding to cosmological distances, it is possible to formulate a relativistic description of the Universe which is consistent with current astronomical observations. This description serves to make predictions about the future evolution of our Universe and also to piece together its past history, not to mention the possibility of gaining information about the very birth of the Universe. Apart from the equations of general relativity, this extraordinary theoretical framework, known as the standard cosmological model, is based upon two further important assumptions.

The first assumption is that, on sufficiently large distance scales, the Universe and all its components can be described using a spatially isotropic and homogeneous geometry, which is to say a spatial geometry that does not admit either preferred directions (isotropy) or privileged points (homogeneity).[1] This is equivalent

---

[1] For the sake of completeness it should be stressed that the large-scale distribution of matter, for distances much smaller than the spatial radius of the

to assuming that, at any given time, the spatial sections of the Universe can be described as spaces of uniform curvature (positive, negative or null). This is quite an oversimplification, but nevertheless a useful hypothesis, which allows one to describe the geometry of the Universe in terms of the well-known Robertson–Walker metric, containing only two parameters: the scale factor R, which is in general time-dependent – and which we shall call, for simplicity (but somewhat improperly), the radius of the spatial part of the Universe – and the uniform curvature of the spatial sections of the Universe.

The other assumption concerns the physical properties of the particles and macroscopic bodies populating our Universe. It is assumed that, on large scales, they behave as a perfect gas with two main components: radiation, whose pressure is exactly equal to one third of its energy density, and non-relativistic matter, with zero pressure. (Recently, it has been found that, apart from these two components, other cosmic components seem to play a fundamental role, as discussed further in Chap. 9.) Moreover, the radiation is assumed to be in thermal equilibrium, i.e., characterized by a black-body frequency distribution (exactly like the electromagnetic waves present in a hot oven), and to evolve in time adiabatically, in such a way as to keep the entropy constant, in agreement with the laws of classical thermodynamics. The presence of this background of cosmic radiation is interpreted as a relic of the Big Bang, i.e., the big explosion from which the Universe and space-time itself were born.

Exploiting these assumptions (which may appear rather an oversimplification, but which are supported by various direct and indirect observations), one can exactly solve the Einstein gravitational equations, thus determining the evolution of the cosmic geometry and its gravitational sources. One then finds that the energy density of non-relativistic matter is inversely proportional to the spatial volume, i.e., to the third power of the spatial radius R. For its part, the radiation energy density is inversely proportional to the fourth power of the radius. Thus, in an expanding Universe where R grows in time, the radiation density decreases in

---

Universe, seem to have a non-homogeneous, fractal-like distribution [F. Sylos Labini, M. Montuori, and L. Pietronero: Phys. Rep. **293**, 61 (1998)]. It is not yet clear at which precise scale homogeneity becomes an acceptable assumption.

time faster than the matter density. Indeed, current cosmological observations indicate that the radiation energy density is today about ten thousand times smaller (i.e., a factor of 10^{-4} smaller) than the matter density.

Going backward in time, however, the energy density of radiation tends to increase with respect to that of matter until, at the so-called equality time, the two energy densities are equal. For times before the equality time, the energy density of radiation takes over, and the Universe undergoes a phase transition where its rate of expansion also undergoes a change.

Hence, according to the standard model, the Universe is characterized by two main stages: an initial phase, where radiation is the dominant form of energy filling the Universe and controlling its evolution, followed by a phase (possibly extended in time until the present epoch) where matter is dominant. The expansion rate of the space-time geometry (more precisely, the speed measuring the rate of change of R in time) is higher during the radiation phase, despite the fact that the energy density, and hence the mutual gravitational attraction, tends to decrease as a function of time. In other words the equations of general relativity provide us with solutions describing a *decelerated* expansion, for both phases. This means, strictly speaking, that in these solutions the first derivative of the spatial radius $R(t)$ is positive, while the second derivative is negative. Furthermore, during both phases the spatial curvature decreases uniformly as the inverse square of the spatial radius; the global, space-time curvature – which is closely linked to the expansion rate – also decreases continuously as a result of the decelerated expansion of R, but at a faster rate than the spatial curvature.

Quite independently from their particular kinematic properties, it should be stressed that the possible existence of cosmological solutions describing an expanding Universe is one of the greatest achievements of the standard model. In fact, one can explain the famous Hubble law (discovered at the beginning of the twentieth century), according to which the light we receive from the various galaxies populating the Universe is characterized by a redshift that is an increasing function of the distance at which the emitting galaxy is located.

In fact, let us consider a light ray currently received on the Earth, and emitted from a galaxy many light-years away. Light propagates with a large but finite speed, so that the light ray has spent many years traveling in space to cover the distance separating us from that galaxy, from which it was emitted long ago. On the other hand, if the Universe is expanding, its radius was smaller in the past than its radius today, the cosmic energy density was more compressed, and the cosmological gravitational field more intense (so that the space-time geometry was more curved than today). Thus, as previously discussed, such a light ray was emitted with a redshift (with respect to the flat space) which is greater than the one affecting light today. This is the reason why the light received from a distant source is redshifted with respect to the same light emitted today, or to the light emitted by a source located at a smaller distance.

Clearly, the redshift depends on the rate of change of the radius of the Universe during the intergalactic journey of the light ray, and thus on how far away the observed galaxy is. The exact relationship between redshift and distance is quite complicated in general, but to first order it can be approximated by a linear relationship. In this case the shift computed according to the standard model turns out to be directly proportional to the distance of the source, just as suggested by the observations leading to the Hubble law, and the proportionality constant H between redshift and distance is called the Hubble constant (or better, the Hubble parameter). In the standard model the Hubble parameter is just determined by the speed at which the spatial radius of the Universe changes with time, i.e., by the first derivative of the spatial radius divided by the spatial radius itself, and is also proportional to the space-time curvature (in agreement with the fact that the reddening of light is a gravitational effect).

Within the physics of electromagnetic waves, on the other hand, the reddening of light is not a new phenomenon. It is well known, even in a non-relativistic context, that the frequency of a periodic signal emitted by a moving source is shifted towards the red or towards the blue, depending on whether the source is receding or approaching the observer, respectively, according to the well-known Doppler effect.

The cosmological redshift described by the Hubble law can also be interpreted as a Doppler effect, associated with the fact that galaxies are mutually receding as a result of the expansion of the Universe. According to this interpretation, the redshift z (given to a first approximation by the product of the constant H and the distance of the source, divided by the speed of light c) also coincides with the ratio between the recession velocity of a galaxy and the speed of light. The recession velocity increases with the distance, and when the distance approaches the characteristic scale c/H, it approaches the speed of light. This is the reason why the parameter c/H (proportional to the inverse of the Hubble constant, and called the Hubble radius, or Hubble horizon) defines a limiting distance scale within which exchange of signals and, consequently, causal interactions are possible.[2]

It should be noted for subsequent applications that, since the parameter H is proportional to the space-time curvature, the horizon radius c/H is inversely proportional to the curvature. Thus, in a time-dependent geometry, the size of the horizon becomes large when the curvature decreases, while it shrinks when the curvature increases. We will come back to this point in Chap. 5, when we discuss the properties of inflationary cosmological models.

Let us now reconsider the solutions of the Einstein equation resulting from the assumptions of the standard cosmological model. Using these solutions we can make predictions about the future evolution of the Universe. It turns out that such evolution is completely determined once we know the present value of the Hubble parameter, the energy density, and the equation of state, i.e., the relation between the effective pressure and energy density, of the dominant gravitational sources. According to the simplest version of the standard model, the Universe should currently be matter-dominated, i.e., filled with a dust-like fluid, whose average pressure is zero. In this case, depending on the value of its energy density, there are two possible classes of future evolution.

If the density is below some critical threshold (which depends on the current value of the Hubble parameter, the speed of light, and Newton's gravitational constant), the three-dimensional space must then have a constant, negative curvature. (This space,

---

[2] It should be noted that, in the context of modern astrophysical observations, it is quite usual to find distant sources characterized by a redshift parameter z larger than one. This does not imply that their recession velocity is larger than the speed of light, however, since the relationship between redshift and velocity, like the one between redshift and distance, is no longer linear for $z > 1$.

which is said to have hyperbolic geometry, is difficult to visualize. Roughly speaking, it is rather like the central cavity of a saddle.) In this case there is an indefinite and gradual slowing down of the expansion rate, until the space-time becomes completely empty and flat. On the other hand, if the density is greater than this critical value, then the three-dimensional space must have a constant positive curvature (like a sphere). The expansion will decelerate until it eventually stops, whereupon the Universe will subsequently start contracting until it collapses into a final singularity. The limiting case in which the density is exactly equal to the critical value corresponds to a zero spatial curvature, and once again to an endlessly decelerated expansion.

Current effort in astronomy and astrophysics thus focuses on the possibility of measuring, either directly or indirectly, the values of the Hubble parameter and the total energy density of the present Universe. The most recent measurements seem to indicate that the spatial curvature is almost zero, in agreement with the predictions of inflationary models (see Chap. 5), and that the total energy density is very close to the critical value. However, oddly enough, the present Universe seems to be in a state of *accelerated* expansion (see in particular Chap. 9).

These observational results, which are receiving more support all the time, clash with the cosmological solutions of the standard model, which do not actually predict any acceleration. They may therefore suggest that the current Universe is *not* matter-dominated, i.e., it may not be filled with a fluid with zero pressure. Instead, its energy density may be dominated by some exotic element dubbed quintessence (as a reminder of the mysterious fifth element of post-Aristotelian philosophy). In order to induce the observed acceleration, this substance should have a nonzero *negative* pressure.

The simplest example of a source that could reproduce quintessence effects is certainly the famous cosmological constant, a term representing the vacuum energy density and introduced into the equations of general relativity by Einstein himself (although he publicly withdrew it, saying that it was the biggest blunder of his life!). Today, it would appear that Einstein was right after all. It should be stressed, however, that the value of the cosmological constant needed to reproduce the observed cosmic acceleration has never received a satisfactory explanation, not even in the context of modern quantum theories which unify all interactions. We will discuss this issue in more detail in Chap. 9.

Besides predicting the future, the standard model allows us to go backwards in time through the history of the Universe, thus reconstructing its past. Using current observations one can compute, for instance, the temperature at which the energy densities of matter and radiation were of the same order. Since the radiation is in thermal equilibrium (as happens in an oven when it reaches its designated temperature), its energy density is proportional to the fourth power of the temperature, according to the well-known Stefan law of classical thermodynamics. On the other hand, according to the equations of the standard cosmological model, the energy density of radiation must also be inversely proportional to the fourth power of the spatial radius of the Universe, as mentioned previously. It follows that the radiation temperature is inversely proportional to the spatial radius, and consequently, as the Universe expands (i.e., as its radius increases), the temperature decreases and the radiation gets colder.

This decrease in temperature is a gravitational effect quite similar to the redshift of frequencies. In the past, at the time of equality between matter and radiation energy density, the Universe was much hotter than it is today. But how much hotter, precisely?

We know that the current radiation temperature is about three degrees kelvin above absolute zero (i.e., about -270 degrees centigrade, or 10^{-4} eV, where the symbol eV denotes electron-volts, a typical unit of energy and temperature used in nuclear physics). We also know, as already stressed, that the current value of the ratio between the radiation energy density and the matter density is about 10^{-4}. According to the standard model this ratio varies with time in a manner inversely proportional to the spatial radius, and it is therefore directly proportional to the temperature (see the previous paragraphs). The equality epoch, occurring when the ratio of the energies was about 10^4 times bigger than its current value, thus corresponds to a radiation temperature 10^4 times higher than the current value, i.e., about one electron-volt (or ten thousand degrees kelvin). Similar arguments can also be applied to compute the temperature at earlier epochs.

We find therefore that, going backward in time according to the standard model, the Universe becomes not only denser and more warped, but also hotter. Cosmic history can then be traced back in terms of three possible evolution parameters: time, temperature, and space-time curvature (or, equivalently, the inverse of the curvature, the Hubble radius c/H). Obviously, these parameters are not mutually independent: time is proportional to the Hubble radius, while the temperature is inversely proportional to the spatial radius R which, in turn, depends on time.

The main stages of the standard cosmological model as a function of the above-mentioned evolution parameters are shown in Fig. 2.1. In this figure one proceeds backward in time from the current value of the Hubble radius down to the Planck radius, i.e., from the current epoch where the Universe has a curvature radius of about 10^{28} cm, or 10 billion light-years (the current value of the Hubble radius), until the beginning of the quantum gravity epoch, when the curvature radius was about 10^{-33} cm (corresponding to the so-called Planck length L_P).

Let us follow, for instance, the scale corresponding to the temperature of the cosmic microwave radiation, and proceed backward in time, starting from its current value of about 10^{-4} electron-volts. When the temperature is about one hundredth of an electron-volt, we reach the epoch of galaxy formation; as the temperature reaches the value of about one electron-volt, we reach the epoch of matter–radiation equality, corresponding roughly to the phase in which nuclei and electrons tend to combine into atoms. When the temperature grows to about 1 million electron-volts (1 MeV), we reach the epoch in which the primordial synthesis of the elements (or nucleosynthesis) was taking place. Neglecting many other processes, for the sake of brevity, we then get to a temperature of about 10^{16} GeV (1 GeV = 1 billion electron-volts), corresponding to the epoch where the electromagnetic and nuclear (weak and strong) interactions were probably unified into a single fundamental interaction. Finally, at a temperature of about 10^{19} GeV (the Planck temperature), we reach the beginning of the quantum gravity epoch where general relativity, together with the standard cosmological model, can no longer be unambiguously applied.

It should be stressed that, for graphical reasons, the different scales used in Fig. 2.1 do not respect the relative lengths of the

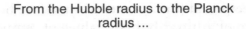

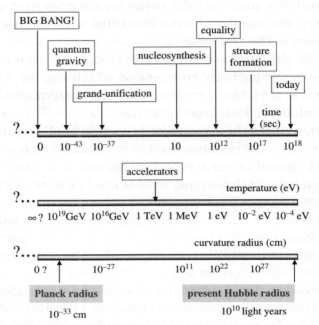

FIGURE 2.1 Simplified sketch of the evolution of the Universe according to the standard cosmological model. The figure summarizes the main stages of cosmic evolution, starting from the present epoch characterized by the observed value of the Hubble radius, back in time down to the quantum gravity epoch characterized by the Planck radius. The past history of the Universe is parametrized in terms of three equivalent cosmological variables: the cosmic time (measured in seconds), the radiation temperature (measured in electron-volts), and the space-time curvature radius (measured in centimeters)

various epochs. However, one can nevertheless ascertain from the figure that the maximum energy scale currently attainable in a laboratory supplied with the most powerful available accelerators (about 1 TeV, i.e., one thousand GeV) is still well below the typical energies coming into play in the primordial Universe.

Tracing the evolution predicted by the standard model backward in time even beyond the Planck radius, the Universe (as shown in Fig. 2.1) necessarily reaches a singular stage where the temperature becomes infinite, the curvature radius c/H is then zero, and its reciprocal (the curvature) therefore becomes infinite. The occurrence of arbitrarily high values of temperature, density and curvature has suggested the name Big Bang for such a singular phase, which the standard model identifies with the beginning of the expansion of the Universe.

If we fix the origin of the temporal axis (i.e., the coordinate $t = 0$) to coincide with the moment at which the Big Bang takes place (as in Fig. 2.1), the current epoch then corresponds to a time coordinate of about 10 billion years (i.e., about 10^{18} seconds). This number, often called the age of the Universe, actually represents the time interval elapsed between the Big Bang and the present epoch, as is clearly shown in the figure. It also coincides with the age of the Universe, strictly speaking, only if the Universe did not exist before the Big Bang. Otherwise, it simply represents the duration of the current phase of the Universe, i.e., of the epoch described by the standard model.

The origin of the time axis is obviously arbitrary, and we could have chosen to set the coordinate $t = 0$ at any other point in the graph. However, what is not arbitrary is the fact that in the standard model the Universe, at some point of its past evolution, necessarily reaches a singular stage where the temperature and the curvature become infinite. Beyond this point the time coordinate *cannot be further extended*, since the presence of a singularity makes any physical model meaningless.

The following question therefore naturally arises at this point: Does the singularity really mark the origin, the birth of the Universe and the beginning of space-time itself, or is it only a shortcoming of the standard model which could be removed within a more detailed, realistic and complete cosmological framework? This is why there are dots and question marks in Fig. 2.1, at the beginning of the various scales. The presence of a singularity is a first, important reason that may suggest the possibility of modifying the standard cosmological model near the initial time. This is not the only reason, however. As will be shown in Chap. 5, there are also other kinematical issues and shortcomings. All of them can be sorted out, at least in principle, provided that the primordial evolution is modified by assuming that the initial Universe was not dominated by radiation (which would dominate only later), and that the initial expansion, contrary to the prediction of the standard model, is not decelerated. Rather, the initial phase of the Universe should be characterized by an accelerated expansion (also said to be inflationary) which only afterwards decelerates, eventually reducing to the standard one.

String cosmology suggests that this inflationary phase preceding the standard one could be identified with the pre-Big-Bang phase introduced in the previous chapter. In the following chapters we shall therefore analyze the physical properties of this phase in more detail, starting from the motivations that would lead us to introduce it into the framework of a cosmological scenario inspired by string theory.

3. String Theory, Duality, and the Primordial Universe

For anybody reading the introduction, there will no doubt have been important questions which arise spontaneously and are left unanswered. For instance, why does what we have called the "pre-Big Bang scenario" emerge within string theory and not within the classical cosmological setup based upon Einstein's equations? And what is string theory?

For the answer to the second question we refer the reader to Chap. 4. With regard to the first question, there are many reasons (that we will discuss below), but it is probably appropriate to say that the fundamental argument pertains to the particular symmetries that are present in string theory and not in Einstein's theory of gravity.

So let us consider the theory of general relativity. As is typical of classical physics, this theory enjoys the following key property of symmetry: any (fundamental, elementary) process described by such a theory is invariant when the sign of the time coordinate is changed (provided this is not in contradiction with the laws of relativistic causality, of course). This is the so-called time-reversal, or time-reflection, symmetry. It implies for instance that, if the equations of that theory admit a solution describing a particle moving at constant speed from left to right, then there must exist a solution describing the same process seen backward in time, i.e., describing an identical particle moving with the same speed but going from right to left.

Furthermore, for a given solution describing a decelerating particle which moves from left to right, there must exist a solution that describes the same particle accelerating in a motion from right to left. In other words, the theory should work like a video tape which allows us to play the recorded images both forward and backward.

It is worth stressing, in particular, that some vectorial quantities, such as velocity, are reversed in sign when time is reversed, while others, like the acceleration, remain unchanged. In fact, if we consider the previous example, we may note that the acceleration is always pointing from right to left. However, in the first case (i.e., when the motion is from left to right) the acceleration is opposite to the direction of the motion, and thus decreases the speed of the particle, while in the second case (i.e., from right to left) the acceleration is pointing along the direction of the motion, and thus increases the speed of the particle.

Within classical cosmology, the time-reversal symmetry implies that, to any given cosmological solution describing an expanding Universe (with a growing spatial radius R, like ours), there must be an associated solution describing the same Universe but evolving backward in time, i.e., a contracting Universe (with decreasing R). Also, if the expanding Universe is decelerated then the contracting (time-reversed) universe will be accelerated, just as pointed out in our previous example regarding the motion of a particle (see Fig. 3.1). Since the behavior of the space-time curvature follows the absolute value of the speed measuring the rate of

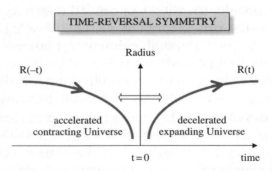

FIGURE 3.1 A pictorial representation of time-reversal, or time-reflection, symmetry. The curve plotted in the positive range of the time coordinate (i.e., to the right of the origin $t = 0$) represents the spatial radius $R(t)$ of a Universe which is expanding (as the radius is growing), but in which the expansion is decelerating (as shown by the concavity of the curve), and which is in a state of decreasing curvature, as the rate of change of $R(t)$ (not plotted in the picture) is decreasing. By applying a time-reflection transformation, $t \to -t$, one obtains the curve $R(-t)$ plotted in the negative-time range (i.e., to the left of the origin $t = 0$). Thus, time reflection transforms expansion into contraction (and vice versa), deceleration into acceleration (and vice versa), and decreasing curvature into increasing curvature (and vice versa)

change of $R(t)$ (see Chap. 2), we may also say that a time-reversal transformation maps expansion into contraction and decreasing curvature into increasing curvature.

It should be stressed that the existence of a possible solution of the cosmological equations does not automatically guarantee that the corresponding scenario does actually occur in nature. In other words, we cannot conclude that, given the existence of our expanding Universe, a corresponding contracting Universe also exists; similarly, we cannot establish that for a given particle moving from left to right we would observe a particle moving from right to left. According to the theory this a possibility, but its occurrence is not mandatory.

General relativity is a classical theory of gravitation, based upon macroscopic observations (Newton's law, the motion of planets around the Sun, and so on), and implicitly rooted in the fundamental concepts of classical relativistic mechanics, generalized to the case of a curved space-time framework. Years of study and joint effort by many research groups around the world have shown that this theory is unlikely to be compatible with quantum mechanics, i.e., with the theory which lies at the heart of a physical description of the microscopic world. This is the reason why general relativity, as a theory of the gravitational forces, has always strongly resisted any attempt to unify gravity with theories describing the other forces active in a microscopic context – namely, the theories describing nuclear (weak and strong) interactions, and the electromagnetic interactions.

A key step forward in this direction does seem possible, however, within the framework of string theory, which should provide a unified description of all the forces of nature, valid at all energies, and including gravity even in its quantum regime. The description of nature proposed by string theory, rather than being based upon point-like elementary objects (the well-known particles of classical physics), admits as building blocks objects that have a spatial extension, albeit one-dimensional. Such objects can be either closed (in the case in which both ends coincide), or open, although sometimes with their ends fixed on some preferred spatial hyperplane: we may visualize them as ordinary strings of finite length and negligible thickness. Different vibrational modes of these strings may simulate the various types of particles and the fundamental interactions that we are currently able to observe.

A more explicit illustration of string theory will be given in the next chapter and in Chap. 10. As far as this chapter is concerned, we just need to point out that according to string theory Einstein's gravitational equations ought to be generalized, and that such a modification brings about two important consequences.

The first is that the gravitational equations provided by string theory, besides the time-reversal symmetry, show another important kind of symmetry, dubbed duality. This symmetry has no counterpart in any type of classical or quantum theory of fields, since (as will become clearer in the next chapter) it is rooted in the fact that the fundamental objects of the theory are extended rather than point-like.

There are, in general, various types of duality symmetry (T, S, U duality), corresponding to different types of transformation that leave the form of the string theory equations unchanged. For instance, T-duality states that if the equations of the theory admits solutions describing universes of radius R, then universes with the reciprocal radius $1/R$ are also possible solutions of the same theory. Similarly, according to S-duality, if the theory admit solutions describing a particle characterized by a charge of strength Q (not necessarily electric charge), then there must be solutions describing a particle with charge $1/Q$. Finally, U-duality is (roughly speaking) a combination of T and S duality.[1]

It is worth noting that a large value of the charge Q corresponds to a small value of $1/Q$ and vice versa; a large value of the radius R corresponds to a small value of $1/R$ and vice versa. In other words, duality relates large universes to small universes, and strong couplings to weak ones. Furthermore, it transforms expansion into contraction, and vice versa (see Fig. 3.2). Indeed, if the function $R(t)$ grows in time, then a Universe of radius R expands, while its dual partner of radius $1/R$ contracts, as $1/R(t)$ decreases.

---

[1] Note that we are working in an appropriate system of units (naturally fixed by string theory, as discussed in Chap. 4), where radius and charge are dimensionless quantities. Otherwise, the reciprocal radius does not have dimensions of length, and one has to restore the correct dimensionality by multiplying $1/R$ by the square of the string length parameter. Similar arguments also apply to the reciprocal of the charge.

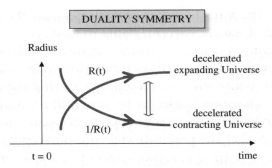

FIGURE 3.2 A duality transformation that inverts the spatial radius, $R \rightarrow 1/R$, transforms decelerated expansion into decelerated contraction and vice versa. The time behavior of the curvature, as well as the acceleration/deceleration properties of the original solution, are left unchanged by a duality transformation

However, since there is no reflection of the time coordinate, there is no change in the acceleration properties of the transformed solution. Thus, decelerated expansion transforms into decelerated contraction. Similarly, the time behavior of the absolute value of the rate of change of $R(t)$ is invariant under the inversion of the radius. Thus, the decreasing-curvature (or growing-curvature) status of the original solution is preserved by a duality transformation.

The second important property of string theory that we would like to recall here is that the implementation of the duality symmetry necessarily requires the introduction of a new type of force into the gravitational equations, mediated by a neutral scalar particle (i.e., by a particle without electric charge and without intrinsic angular momentum). This particle is called a dilaton, and under a duality transformation the force field associated with the dilaton is not invariant in general, whence different, duality-related solutions of the theory are characterized by different values of the dilatonic forces.

Within string theory, the gravitational importance of the dilaton is encoded into the fact that this field determines the value of the effective Newtonian constant G which, in its turn, fixes the strength of the gravitational interaction (as will be illustrated in Chap. 4). Applying a duality transformation that changes the value of the dilaton it is then possible to modify the value of the effective gravitational coupling. In such a context, the Newtonian constant loses the role of fundamental gravitational parameter, and the theory may describe physical situations where the gravitational force can be either weaker or stronger than we usually experience. Furthermore, the strength of the interaction may not be constant, but vary in space and time, following the dilaton behavior.

This is certainly a big physical revolution introduced by string theory and, as we shall see, it may have important consequences in a cosmological context. It is true that the possibility of a variable gravitational constant, associated with the presence of a scalar field in the gravitational equations, was previously suggested by Carl Brans and Robert Dicke in 1961 (well before string theory was introduced). However, it is only through string theory that such a variable-G scenario finds robust motivations, and actually becomes essential for the consistent formulation of a quantum theory of gravity.

In a cosmological context, the simultaneous implementation of the duality symmetry and the time-reversal symmetry allows us to obtain new cosmological solutions (i.e., models for the Universe) that were not contemplated by general relativity, as suggested by the theoretical physicist Gabriele Veneziano in 1991 and later elaborated by him, in collaboration with the present author.

Let us consider, for instance, the current Universe, assuming that it can be properly described by the solutions of the standard cosmological theory, and let us focus our attention on an expanding spatial section of spherical shape. The corresponding spatial radius $R(t)$ increases with time, while the space-time curvature, being proportional to the square of the expansion velocity, decreases with time, as the expansion rate is slowing down according to the standard cosmological scenario. Let us then apply a time-reflection transformation to this solution, that is, let us reverse the time arrow. As previously pointed out, we will obtain a new solution in which the curvature is increasing, and the transformed Universe will contract, as the radius $R(-t)$ is a decreasing function of time. Finally, let us apply a duality transformation, and invert the radius: the curvature will keep increasing, while the Universe will become an expanding universe, since if $R(-t)$ is decreasing in time, its reciprocal $1/R(-t)$ is increasing (see Fig. 3.3).

We are thus led to the following result. Thanks to the combined action of duality symmetry and time-reversal symmetry,

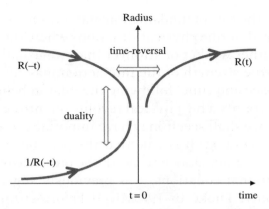

FIGURE 3.3 Pictorial representation of the combined action of a time reversal and duality transformation applied to a given initial solution $R(t)$ (*top right curve*), describing the decelerated expanding radius of a standard (decreasing-curvature) model of the Universe. The final result is the curve $1/R(-t)$, plotted in the *bottom left* of the figure, describing accelerated expansion and growing curvature

with any cosmological solution describing an expanding Universe with decreasing curvature we can associate another possible solution describing an expanding Universe with increasing curvature. In particular, if the two solutions are smoothly matched at the time $t = 0$, one obtains a cosmological model where the spatial radius of the Universe increases continuously from zero to infinity, as illustrated in Fig. 3.4 (top panel). Let us compute the space-time curvature corresponding to this model, namely the absolute value of the speed measuring the rate of change of R (more precisely, the absolute value of the time derivative of the radius divided by the radius itself). The resulting plot will grow in the range of negative values of the time coordinate (left of the origin), and decrease in the positive range (right of the origin). Assuming that the standard solution and its dual partner join continuously across the origin, $t = 0$, we then find the characteristic bell-shaped behavior for the curvature, as shown in Fig. 3.4 (bottom panel) and anticipated in Chap. 1 as typical of the pre-Big-Bang scenario.

We should recall, also, that the time derivative of the radius divided by the radius itself defines the Hubble parameter H introduced in the last chapter. We can say, therefore, that duality and time-reversal transformations map an expansion with decreasing

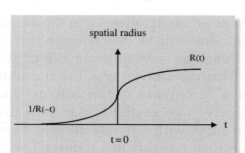

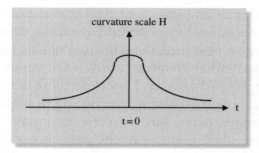

FIGURE 3.4 Time evolution of the spatial radius (*top panel*) and the space-time curvature (*bottom panel*), in a cosmological model based upon the duality symmetry of string theory. The curvature is parametrized by the absolute value of the Hubble parameter H, defined as the time derivative of the radius divided by the radius itself. The standard cosmological phase is described by the curves plotted in the positive range of the time coordinate, i.e., to the right of the origin $t = 0$. The complementary pre-Big-Bang phase, plotted to the left of $t = 0$, is obtained by applying a duality and a time-reversal transformation to the standard solution

H, typical of the standard cosmological phase, into an expansion with increasing H, typical of the pre-Big-Bang phase. Also, and most importantly, the decelerated expansion of the standard solution is mapped into an accelerated expansion of the transformed solution. This is a consequence of time-reversal symmetry, as already outlined in the example of a moving particle. (The relevance of the fact that the dynamics of the pre-Big-Bang phase becomes accelerated will be explained in Chap. 5.)

As we have already stressed, implementation of the duality symmetry requires inversion of the radius to be accompanied by a simultaneous transformation of the dilaton field, according to the rules dictated by string theory. Since it is the dilaton which fixes the value of Newton's constant, it follows that the duality-transformed cosmological solutions will be characterized by different values of G, namely by different intensities of the corresponding gravitational forces. This certainly makes the transformed solutions unacceptable for a description of the current Universe. Indeed, high-precision observations establish that the current value of G is nearly fixed in time, with possibly allowed annual percentage variations smaller than one part in a thousand billion.[2] However, this constraint does not prevent that in the remote past; before the formation of galaxies and stars, and even before atomic nuclei were formed, the gravitational force could have had a different value.

This may be a first hint that the dual of the standard cosmological solutions, rather than the current Universe, may describe the Universe in its early stages. However, there is more. There are also thermodynamical arguments according to which, by exploiting the duality symmetry, our Universe may have undergone an epoch with "specular" features in the past, as compared with the current one. This possibility was put forward at the end of 1980s by the pioneering work of some cosmologists and string-theory experts like Robert Brandenberger and Cumrum Vafa, followed by later work by the theoretical physicist Arcady Tseytlin (see also the string gas cosmological scenario discussed in the next chapter).

After this discussion, it is probably more evident to the reader why the existence of cosmological solutions that expand with increasing curvature is a peculiar property of string theory: the appearance of these solutions, as well as their close link to the solutions of the standard scenario, is a direct consequence of duality, namely of a new "stringy" symmetry which is absent in the pure Einstein theory. Obviously, as already pointed out, the existence of allowed solutions of a theory does not necessarily imply that the scenario they describe actually occurs in nature. However, the presence of these dual solutions suggests a possible answer to one of the key questions implicit in standard cosmology (as well as in inflationary cosmology, discussed in Chap. 5), i.e., to the question of the initial state of the Universe (assuming that the singularity is somehow avoided).

---

[2] See for instance J.D. Barrow: Phil. Trans. Roy. Soc. Lon. A **363**, 2139 (2005).

String theory suggests that, initially, the state of our Universe might correspond to the state determined by applying a duality and a time-reversal transformation to the current cosmological state. The cosmological scenario that completes the standard evolution by adding the phase dubbed "pre-Big-Bang", introduced in Chap. 1, does indeed emerge from the assumption that the evolution of the Universe should be self-dual and time-symmetric, i.e., simultaneously invariant under the combined action of duality and time-reversal transformations (as in the case of the cosmological model shown in Fig. 3.4).

Under such a hypothesis the current Universe, which is characterized by an almost flat space-time geometry and by an average energy density and temperature much lower than their standard macroscopic values, should have had, in its very early past, a dual counterpart similar to its present state. Hence, the Universe should have undergone a very early regime associated with an almost flat, empty and cold state which, going backward in time, gets progressively more and more flat and empty until it corresponds, asymptotically, to the state called the perturbative vacuum of string theory. In the primordial cosmological phase, that we now identify with the epoch preceding the Big Bang, the growth of the curvature has led the Universe towards progressively more curved and denser states, until the radiation produced at a microscopic level became dominant, causing the primordial explosion that finally led to the current (standard) decreasing-curvature regime.

For a true self-dual scenario, however, it is essential that the evolution of the geometry is accompanied by the evolution of the dilaton field, which is present in all string theory models. In the current cosmological state the gravitational force has a nearly constant strength, controlled by the Newtonian constant G; hence the dilaton, which fixes this strength, must be constant. In the primordial state representing the dual counterpart of the present one, it turns out – according to the rules of the duality transformations – that the dilaton has to increase with time, thus describing an increasing gravitational coupling. It follows that the current value of the Newtonian constant is reached after starting from an almost zero initial asymptotic value, as illustrated in Fig. 3.5 (right panel).

This feature of the dual solutions has an important physical consequence. Indeed, in all unified models based upon string

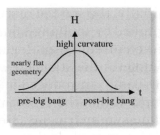
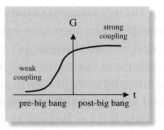

curvature scale H   Newton constant G

FIGURE 3.5 Time evolution of the curvature (represented by the Hubble parameter H) and the Newtonian constant G (determined by the dilaton), for a typical self-dual solution of the string cosmology equations. The initial cosmological configuration – approaching the so-called string perturbative vacuum at very large negative times – is characterized by a nearly flat space-time geometry and the vanishingly small intensity of all interactions

theory – i.e., in all models which, using strings, try to incorporate a somehow unified description of all forces of nature – the dilaton must determine not only the value of G, but also the coupling strengths of the other fundamental forces (see Chap. 4). Hence, in the context of a self-dual cosmological scenario, one finds that, in the initial state of the Universe, all the coupling parameters tend to approach zero. In other words, all the forces and all the interactions are asymptotically suppressed, following the behavior of G shown in Fig. 3.5. This implies that the initial evolution of the Universe, starting from the asymptotic state called the perturbative vacuum, can be correctly described using the string equations in the semi-classical limit and weak-coupling regime; the quantum (higher-order) corrections are expected to become relevant only later when, as the time of the Big Bang gets closer (i.e., around $t = 0$), the various forces become sufficiently strong.

It should be stressed at this point that, if the self-dual symmetry of the cosmological evolution were to be exact, then any instant in the life of the current Universe would have its dual counterpart in the past. However, we have learned many times in the history of physics that the symmetries of our theories, when implemented in nature, are not always exact. Frequently, they are only valid in some approximate limit, since there are physical effects responsible for their breaking (through a spontaneous, or more involved, mechanism). As far as cosmology is concerned, we know in advance that the duality symmetry cannot be exactly implemented at all times. Indeed, the current state of the Universe contains matter and thermal radiation characterized by a high entropy level, contrary to the asymptotically empty and cold initial state. This means that, in a realistic cosmological scenario, the behavior of the space-time curvature will not be represented by an exactly symmetric curve (like the ones shown in Figs. 3.4 and 3.5); rather, it will have some distortions caused by an approximate implementation of duality and time-reversal symmetry. This by no means contradicts the basic idea suggested by duality, i.e., the fact that our Universe could have undergone in the past an accelerated phase with increasing curvature, complementary to the current one.

Within modern theoretical physics, duality is a quite important symmetry. There is a common belief that the duality symmetry is somehow present in nature at a fundamental physical level. The pre-Big-Bang scenario, based upon this symmetry, is thus supported by rather solid theoretical motivations. It should be recalled, in fact, that the practice of exploiting the symmetries of the theory in order to formulate predictions about as yet unobserved phenomena, is a common working method employed by physicists, a method which in the past has led to many important discoveries. It is probably appropriate to draw an analogy here with another important symmetry of modern theoretical physics, known as supersymmetry.

According to supersymmetry, any physical state with statistical properties of bosonic type (e.g., a particle whose intrinsic angular momentum is measured, in quantum units, by an integer number) must correspond to a supersymmetric "partner" whose statistical properties are of fermionic type (e.g., a particle with half-integer intrinsic angular momentum). Given the presence in nature of known bosonic particles (like the photon, the graviton, and so on), and using supersymmetry, we may then infer the existence of new fermionic particles associated with the previous ones (and suggestively called the photino, the gravitino, and so on), even if such particles have not yet been observed. The validity of supersymmetry as a true symmetry of nature has still to be confirmed experimentally, but its predictions have been taken so seriously that in many laboratories around the world (including CERN, the large European laboratory for particle physics and nuclear research in Geneva), studies are currently being undertaken to detect some supersymmetric partners, using the most powerful particle accelerators presently available.

In a similar fashion, duality symmetry (with the help of time-reversal symmetry) establishes that any expanding geometrical configuration characterized by decreasing curvature corresponds to a dual partner characterized by expansion and increasing curvature. On the other hand, our Universe is currently associated with a post-Big-Bang, decreasing-curvature state. Assuming that the duality symmetry is in fact realized in nature, and in particular (even if approximately) during the course of cosmological evolution, we may then expect the Universe to have undergone in the past a phase characterized by an increasing-curvature expansion.

At this stage, an obvious observation springs to mind. A phase of increasing curvature, if unbounded, could bring the Universe towards a state of infinite curvature, thus introducing the pathology associated with the presence of a singularity, as in the standard model – with the difference here that the singularity, rather than being in the past, would be located in the future. The answer is that, in contrast to models based upon general relativity, such a pathology is not necessarily present in string cosmology thanks to another important feature of string theory, namely, the presence in this theory of a typical fundamental length L_s (see Chap. 4).

The value of this length, at least according to the most conventional string-theoretical schemes unifying all interactions, is expected to be about 10^{-32} cm, i.e., almost an order of magnitude bigger than the so-called Planck length L_P, which characterizes the length scale at which quantum gravitational effects become important. In any case, this length L_s determines the minimum characteristic size for the spatial extension of any physical system, and therefore also for the space-time curvature radius of the Universe, or, equivalently, for the Hubble radius c/H, as these two quantities are proportional (see Chap. 2).

During the initial pre-Big-Bang evolution the Universe, starting from an empty and nearly flat state, becomes progressively more and more curved, so that the absolute value of the Hubble parameter grows in time, while its reciprocal, representing the curvature (or the Hubble) radius, shrinks monotonically. The geometry, on a cosmological scale, may thus evolve towards the state corresponding to a curvature radius of the same order as L_s. We may expect, however, that this threshold cannot be crossed, i.e., that the growth of the curvature has to stop when it reaches the string scale, since a greater value of the curvature would correspond to a Hubble radius smaller than L_s, and this seems to be meaningless for a model based upon strings. Hence, after reaching this threshold configuration, the curvature should either remain constant or start decreasing.

The detailed mechanism by which the increasing curvature is tamed – smoothing out the singularity, and eventually decreasing according to standard model behavior – is not yet fully understood, mainly due to technical issues related to the presence of quantum effects (and higher-order string theory corrections) near the maximum curvature regime (see Chap. 8). We may recall in particular that, when the curvature radius of the Universe reaches L_s, the gravitational equations are drastically modified: an infinite sum of terms has to be added up, and even the classical notion of space-time becomes inadequate to describe processes that take place in such a regime. However, even in that case, the duality symmetry may play an important role, and this reinforces the key importance of such a symmetry within models of pre-Big-Bang evolution.

However, duality alone cannot account for a consistent formulation of the cosmological scenario described in the previous paragraphs. Another important element concerns the properties of the string perturbative vacuum, the asymptotic state characterized by flat geometry and zero coupling constants, which duality suggests as a possible representation of the initial configuration of our Universe.

If such a state were stable, it could not represent the initial stage of the evolution since the Universe, once in such a configuration, would be eternally trapped there, remaining forever flat, cold, and empty, i.e., radically different from the Universe that we currently observe. Instead, the perturbative vacuum is unstable, i.e., it tends to decay spontaneously exactly like an atom or a molecule which, starting from an excited state, tends to reach another configuration made more favorable by the forces involved. In particular, the perturbative vacuum tends to evolve towards a non-static configuration where the curvature and the dilaton both increase with time, just as expected in the context of a self-dual pre-Big-Bang scenario.

This instability of the initial state is linked to the fact that the expansion of the Universe provides a negative (gravitational) contribution to the total energy of the system, while the increase of a field like the dilaton yields a positive (non-gravitational) contribution. The two energies compensate each other and, as a consequence, the simultaneous increase of the spatial volume (due to the expanding geometry) and the growth of the coupling constants (due to the dilaton evolution) mutually sustain each other, and can be "spontaneously" ignited without any variation of the total energy of the cosmological system, i.e., without feeding this process with external energy sources.

More details about the decay process of the string perturbative vacuum will be provided in Chap. 5. Here we limit ourselves to a somewhat more intuitive description of its instability, considering the motion of some strings within the primordial Universe, and using a simple model where no other matter sources or fields are present than these strings and the dilaton. Hence, the string distribution will be responsible for determining the gravitational field (i.e., the geometry) on a cosmological scale while, at the same time, the behavior of the geometry and the dilaton will determine the evolution of strings and their dynamics.

Let us suppose we want to solve the equations of motion for the full system consisting of gravitational field, strings, and dilaton, taking into account their mutual correlations. Let us impose – as particular initial conditions – that the system starts evolving from a state which, going sufficiently far back in time, tends to coincide with the perturbative vacuum. Finally, let us ask whether our system tends to go back towards the initial, asymptotically free configuration (i.e., towards the perturbative vacuum), or whether it tends to go away from it, beginning a one-way journey towards the high-curvature Big Bang regime. In the former case the perturbative vacuum would be stable, while in the latter it would be unstable.

To answer this question it may be useful to recall that, in the dual solutions describing the state of the Universe before the Big Bang, the cosmological geometry is characterized by an increasing curvature and thus also, as repeatedly stressed (and illustrated in Figs. 3.4 and 3.5), by a monotonic growth of the Hubble parameter H. The reciprocal of the Hubble parameter defines the quantity c/H introduced in Chap. 2, and called the Hubble radius, or Hubble horizon; actually, it represents the spatial size of what is technically called an event horizon, since it measures the maximum distance within which exchange of signals, and consequently causal interactions, are allowed (see Chap. 2). What is relevant in this context is that, if H increases during the regime of pre-Big-Bang evolution, then its reciprocal decreases, so that the Hubble horizon tends to shrink with time.

Within a standard gravitational theory like general relativity, which is based upon classical field theory, and in which the sources of the force fields as well as the fundamental test bodies are point-like objects, the occurrence of a shrinking horizon does not cause any trouble, either practical or conceptual: the number of point particles included within the causal horizon may decrease, but there is no traumatic consequence for the system. However, for a theory in which the fundamental objects are extended, like the strings under consideration, the shrinking of the horizon may lead to a potentially pathological configuration.

Indeed, since the horizon tends to shrink with time, sooner or later a very long string will become greater than the horizon itself, i.e., there will be a situation where parts of the string are within the horizon, while other parts are outside it, without any chance of causal contact, and hence without any possibility of exchanging information between these two parts. It is like a man having his whole body intact but with an invisible barrier, impassable to any signal, cutting the body at the level of the stomach: the head could not know what the feet were doing, and would not even know whether the feet and the legs still existed, and vice versa.

Such a situation, certainly bizarre and somehow unthinkable in the context of classical theories based upon the notion of a point, is possible in a theory containing extended fundamental objects like strings. In particular, when the proper length of a string becomes larger than the horizon, and the propagation of causal signals from one end to the other is blocked, the string is said to be frozen. In that case the string somehow loses its own life, stops oscillating, and starts passively following the evolution of the surrounding geometry, as would happen to a tiny object that was frozen into an ice-block which is part of an iceberg: when the iceberg goes adrift, the object inside the block passively follows its movements.

Even frozen outside the horizon, however, the strings may have an important cosmological effect, since an ensemble of such strings behaves like a gas with negative pressure (as shown by a series of studies initiated by Norma Sanchez, Gabriele Veneziano, and the present author, and later pursued with the collaboration of Massimo Giovannini and Kris Meissner). A negative pressure does not in fact oppose, but tends to favor the increase of both the curvature and the dilaton. As a consequence, it accelerates the shrinking of the horizon, triggering a back-reaction mechanism which renders the initial configuration (the perturbative vacuum) highly unstable.

We may indeed compare the initial state to a small ball on the top of a steep hill: as soon as a breeze moves the ball from its privileged position, it starts rolling towards the bottom of the hill. In a similar fashion, in our simple model of a universe filled with strings, as soon as the curvature starts to increase (for instance as a consequence of unavoidable quantum fluctuations), the horizon starts to shrink, so that more and more strings become larger than the horizon and get frozen, whence their negative pressure brings the Universe even further into the pre-Big-Bang phase, accelerating its race towards states of progressively increasing curvature and shrinking horizon.

Concluding this chapter, we may say that string theory suggests, in various ways, the possibility that our Universe emerges from a primordial state which is unstable, empty, and flat, and has no interactions. The Big Bang, within this scenario, is interpreted as a moment of violent and explosive transition from an increasing-curvature phase to a decreasing-curvature phase, thus corresponding to an intermediate stage in the history of our Universe rather than to the beginning.

If we accept such ideas, at least as a working hypothesis, and take seriously the possibility that the past Universe may have undergone a phase which is (at least approximately) related by duality to the present one, a particular question comes to mind: Is such a phase (so radically different from the standard one) compatible with the subsequent cosmological evolution, i.e., is it possible to join this phase consistently with the standard-model cosmology?

We are asking in particular whether the time evolution of the pre-Big-Bang geometry, despite the great differences with respect to the conventional one, has the right properties to solve the kinematical problems which (as we shall see) are present in the standard model of cosmological evolution.

The answer to those questions will be discussed in the following chapters, after a short interlude devoted to those readers wishing to learn a little more about string theory. In the next chapter, in fact, we will try to give the reader a more precise idea of why a theory of strings may help to avoid the initial singularity, and why such a theory necessarily leads to a modification of the equations of general relativity, providing a different and more complex gravitational theory.

4. The Theory of Strings

This chapter is devoted to those readers interested in the physical foundations of pre-Big-Bang cosmology and who wish to learn the basic concepts of the theory of one-dimensional extended objects – usually dubbed string theory. In particular, we will try to outline here the quantum origin of the duality symmetries mentioned in the previous chapter. Further aspects of string theory pertaining to the unified description of all natural forces will be reviewed in Chap. 10.

The standard model of elementary particle physics – not to be confused with the standard cosmological model describing the Universe – is based upon two milestones of 20th century physics, quantum mechanics and special relativity, and upon the heuristic hypothesis that elementary particles, in the classical limit, are point-like objects without any spatial extension. As shown in a book written by Steven Weinberg,[1] these assumptions lead almost uniquely to the description of elementary particle physics in terms of what is called quantum field theory, a theory based upon the principle that any field strength (for instance the electromagnetic field strength) can always be measured with arbitrary precision at any given point in time and space.

The standard model, built upon those grounds, has achieved a success beyond all possible expectations. Many relevant theoretical predictions of this model have been confirmed experimentally, and some of these predictions are very important, especially those tested by the particle accelerators operating at CERN (Geneva,

---

[1] S. Weinberg: *The Quantum Theory of Fields: Foundations* (Cambridge University Press, Cambridge 1995). The author of this book was awarded the Nobel Prize for Physics in 1979, together with Sheldon Glashow and Abdus Salam, for theoretically predicting the existence of elementary particles associated with the transmission of the so-called weak nuclear interactions. Later, those particles were directly detected at CERN by Carlo Rubbia and Simon Van Der Meer – both awarded the Nobel Prize for Physics for this discovery in 1984.

Switzerland) and FERMILAB (Chicago, USA). Unfortunately, however, the standard model leads to a consistent unified description of only three out of the four fundamental forces of nature, i.e., it unifies the electromagnetic, strong nuclear, and weak nuclear forces, but it does not include the gravitational force – which may be negligible at the typical scales of nuclear and subnuclear physics, but which certainly plays an important role at higher densities and energy scales, like those appearing in a cosmological context. In other words, we can say that the standard model effectively combines quantum mechanics with special relativity, but is unable to combine quantum mechanics with general relativity.

The physical reason why it is so difficult to reconcile quantum mechanics with general relativity is essentially rooted in Heisenberg's well-known uncertainty principle. General relativity is in fact a local field theory like the other theories of the standard model, based upon the assumption that the gravitational field can be measured at any given point in time and space. However, according to the above-mentioned uncertainty principle, an arbitrarily large accuracy in the position implies a full uncertainty in the velocity. In other words, if we measure a gravitational field at a given point with great accuracy, we end up with a correspondingly large uncertainty in its energy. On the other hand, such a large energy fluctuation is necessarily associated with a large fluctuation in the gravitational field itself since – as we have already stressed – it is just the energy which plays the role of gravitational "charge", i.e., source of the gravitational field. Taking into account this intrinsic quantum uncertainty, a gravitational field cannot be measured at a given space-time point with arbitrary accuracy.

For this reason – and only for gravity – an insuperable problem arises when we attempt to combine a local theory like general relativity with quantum mechanics. One may think it would be possible to overcome this problem in various ways, one being the idea that the gravitational force should not be quantized at all (such a drastic alternative would nevertheless require the solution of a number of conceptual and experimental problems). String theory, on the other hand, proposes to overcome this obstacle by abandoning the property of locality, i.e., the requirement that any field should be measurable at any given time and position.

Originally, when string theory was first formulated in the early 1970s, inspired by a model developed by the theoretical physicist Gabriele Veneziano, it was aimed at describing strong nuclear interactions. Later in the 1980s the model was extended to include supersymmetry, and to provide a consistent and compelling framework for a unified theory of all interactions (see Chap. 10). The number of research scientists that have contributed to this project, and are actively working in this field, is too large to mention all of them here. As for this book, it will suffice to observe that this theory is still based upon quantum mechanics and special relativity, but removes the fundamental hypothesis (present in a quantum field theoretical context) that the building blocks of our physical description should have a point-like nature, assuming instead the existence of elementary fundamental objects with a string-like, i.e., one-dimensional structure.

There are two possible kinds of strings: those characterized by free ends (the so-called open strings) and those closed upon themselves (the closed strings). For both configurations the elementary building blocks of the theory are characterized by a finite spatial extension so that, using such objects, the possibility of local measurements of any field (electric, magnetic, gravitational, etc.) at a given space-time point is not only practically impossible, but is also forbidden in principle. Within this framework, all the standard-model problems related to the description of the gravitational field at the quantum level are condemned to disappear.

Furthermore, not only is a theory obtained by replacing points with strings compatible with gravity in the quantum regime, but it also automatically predicts that gravity has to be included among the fundamental forces of nature. In fact, strings are not static entities. Besides their center of mass motion (with an associated translational kinetic energy), strings can vibrate and oscillate as elastic bodies. According to quantum mechanics, however, only a set of discrete values is allowed for the energy and the angular momentum assigned to the various oscillatory states (exactly as happens for the energy levels of an atom). These discrete levels of a vibrating string are associated with a spectrum of states of different masses and angular momenta, describing different elementary particles, just in the same way as the different atomic frequencies are associated with the different spectral lines of the various atomic elements. And here we find the "miracle" connecting strings to gravitational interactions.

In fact, looking at the subset of states describing massless particles, we find that the spectrum of open strings contains a vector field which satisfies all the symmetry properties required to represent an interaction of electromagnetic type. Furthermore, the spectrum of closed strings contains – besides other fields, like the dilaton – a symmetric tensor field which has all the required physical properties of the graviton, representing the "quanta" of the gravitational interaction. On the other hand, closed strings are always (and necessarily) contained in all string models aimed at a unified description of all fundamental interactions (as will be discussed in Chap. 10). It follows that unified models based on strings must *necessarily* encode a tensor interaction of gravitational type, so that the existence of the gravitational force is guaranteed, at both the classical and quantum level. However, the gravitational field equations predicted by string theory are in principle different from the ones predicted by Einstein (the string equations generalize the Einstein equations in a way that will be illustrated at the end of this chapter).

In a similar fashion – and also because a quantum string vibrates in a multidimensional space (with at least nine spatial dimensions, see Chap. 10) – the quantum spectrum of an oscillating string includes other (possibly massive) states, which are appropriate candidates to describe the quanta of strong and weak nuclear forces. All these particles disappear in the limit where the theory becomes purely classical, so they are associated with intrinsic quantum effects. It is exactly this feature that allows string theory to provide (in principle) a quantum description of all known natural forces, without facing the locality problems arising in the context of field theory.

We can say, therefore, that the most relevant features of string theory are linked to the fact that quantum mechanics itself, when applied to extended objects, becomes somehow helpful, instead of giving problems as happens in a conventional field theory based upon point-like objects. Indeed, it is just quantum mechanics that provides the string with a minimum characteristic size L_s (the analogue of the Bohr radius in the case of atomic physics). Thus, while it would be possible at the classical level to conceive of an arbitrarily small string, eventually allowing local measurements of a field, at the quantum level this turns out to be forbidden – exactly in the same way as stable orbits with the electron too close to the nucleus are forbidden in the quantum mechanics of the atom.

At this point, an interesting question comes to mind: How long should these strings be? Their characteristic quantum length L_s represents a new fundamental constant, which can be expressed in terms of the Planck constant $\hbar$, the speed of light c, and the string tension T (i.e., the string energy per unit length, a constant parameter appearing in the analytic formulation of the theory). In principle this length (or, equivalently, the string tension) is an arbitrary parameter – actually, it is the only truly arbitrary parameter present in string theory – so that it can be conveniently tuned to any suitable value determined by the kind of forces we aim to describe with the theory. Several years ago, for instance, when strings were used to build a model of strong nuclear interactions, the value of the fundamental string length was assumed to be of the order of the nuclear radius (about 10^{-13} cm).

Within the context of modern string theory, however, the constant L_s is fixed so that the theory may be able to describe *all* natural forces in a unified fashion. In this scenario, in fact, there are "additional" spatial dimensions that must be added to our three-dimensional macroscopic space in order to implement a consistent quantization of the string motion. Such extra dimensions are certainly not as large as the three ordinary ones, having escaped direct detection up to now: they are supposed to be wrapped (or more precisely compactified), so as to occupy a finite (and possibly small) volume of space, with a size naturally determined by the string length parameter L_s.

On the other hand, the compactification of the extra dimensions to small scales (i.e., the so-called process of dimensional reduction) is closely related to the process which reduces the higher-dimensional (unified) interactions to the standard form of the interactions that we are currently experiencing. To be consistent with the standard-model interactions, in particular, it turns out that the size of the extra volume of space – and thus the string length – has to be tiny. The expected value for L_s is about 10^{-32} cm, i.e., one tenth of the Planck length (barring some "membrane" models that will be discussed in Chap. 10).

It should be stressed that the introduction of the new constant L_s does not increase the number of the fundamental constants in nature. Rather, this number drops drastically. String theory has only two fundamental constants, the speed of light c (which is finite, according to special relativity) and the string length L_s (which is necessarily associated with quantization). In such a context, even the Planck constant itself is a derived quantity. A question then arises: What about all other constants of nature, determining for example the gravitational force, the electrostatic force, and even the size of the hydrogen atom?

The answer to this question highlights another peculiar feature of string theory. In contrast to what happens in the standard model of elementary particles, the fundamental constants of nature cease to be arbitrary numbers determined only by experiment. Instead they are dynamical variables, determined by the expectation values of some fundamental fields – for instance, the already mentioned dilaton – given by the theory. Being expectation values, such constants should be calculable within a given theoretical model, once the current state of the Universe is fixed. However, this procedure, while straightforward in principle, turns out to be difficult to apply in practice to realistic scenarios, owing to computational difficulties.

The most peculiar (and most relevant, for the purpose of this book) example of "promotion" of fundamental constants to dynamical quantities is provided by the dilaton, a new field which is not contained in the standard model, but which is unavoidably present in all string theory models. This field determines the coupling strength of all the fundamental forces, as outlined already during the 1980s by the theoretical physicist Edward Witten – one of the world's leading experts on the theory of superstrings (i.e., string models whose formulation includes both boson and fermion variables, which get interchanged under the so-called supersymmetry transformations). The non-trivial link between the dilaton field and the effective coupling strengths is a typical property of string theory, important enough to deserve at least a short illustrative discussion.

Consider a string, embedded in an external (higer-dimensional) space. Like a particle, its propagation from one spatial position to another describes a continuous trajectory in the external space-time manifold spanned by the whole set of space and time coordinates. However, the trajectory of a point particle is represented by a one-dimensional curve – the so-called world-line of the particle – parametrized by a single time-like coordinate, while the trajectory of the string is represented by a two-dimensional surface – the world-sheet – parametrized by one time-like coordinate describing the evolution in time of the string, and one space-like coordinate describing the spatial positions of the different points of the string at a given instant of time (see Fig. 4.1).

Consider now a closed string, represented by a circle, whose propagation in the space-time manifold describes a cylindrical world-sheet surface. Due to quantum effects (or because of external interactions), the string may split into two strings, which subsequently recombine to form the initial string once again. This process may also occur for a particle in a quantum field theoretical context, and in that case it is represented by a picture – called a one-loop Feynman graph – describing the splitting of the particle world-line. In the string case, the splitting of the cylindrical world-sheet surface will produce a surface with the topology of a torus (namely, a sphere with a hole through it), as illustrated in Fig. 4.2

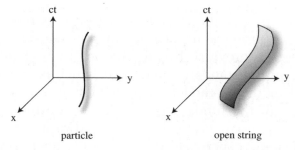

particle   open string

FIGURE 4.1 Time-evolution of a point-like particle (*left*) and an open string (*right*) in the external space-time manifold. The *vertical axis* corresponds to the time-like coordinate, the *horizontal axes* to space-like coordinates. As time goes on the particle moves from one point to another, and the continuous sequence of its (point-like) spatial positions at different times describes a one-dimensional trajectory, the particle world-line (*left*). The string is also moving in space but, at any given time, it is characterized by a one-dimensional (finite) spatial extension. The continuous sequence of the spatial positions of all points of the string describes a two-dimensional trajectory, the string world-sheet (*right*)

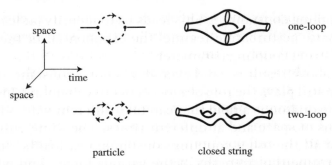

FIGURE 4.2 One-loop (*top*) and two-loop (*bottom*) graphs for a point particle (*left*) and for a closed string (*right*). In the left-hand pictures the world-
line of a physical point particle (*solid curve*) splits into "world-loops"
(*dashed curves*), representing virtual particle–antiparticle pairs generated
by quantum interactions. In the right-hand picture, where the world-lines
are replaced by cylindrical world-sheet surfaces, the same processes are
illustrated for the case of a closed string. Note that in this figure the
time-like axis lies along the horizontal direction, for reasons of graphical
convenience

(top). The same description applies to quantum processes of higher
order, as shown in Fig. 4.2 (bottom) for the two-loop case.

A string process with n loops will be described, in general,
by a two-dimensional world-sheet surface with n "handles", also
called (more technically) a surface of genus n. Quantum interactions among strings can then be approximated by a series of world-
sheet configurations of higher and higher genus. The genus, being
a topologically invariant property of the surface, can be expressed
in terms of the intrinsic curvature of the world-sheet surface (more
precisely, as an integral of the two-dimensional scalar curvature).
The dilaton, by definition, is directly coupled to such a curvature.
In particular, a given dilaton ϕ appears as a multiplicative factor of
the curvature and thus, if it is constant, also of the genus n.

The quantum description of string interactions, on the other
hand, is based on what is known as the partition function, which is
proportional to the exponential of the dilaton–curvature coupling
(in our case, to the exponential of the factor ϕn). An expansion of
the partition function in a series of higher-genus world sheets (i.e.,
a sum of terms with $n = 0$, $n = 1$, $n = 2$, and so on) thus becomes an
expansion in powers of the exponential of the dilaton, exp ϕ. But, by
definition, the loop approximation is an expansion in powers of the

string coupling constant g_s^2. This leads us to identify (at least in the approximate, perturbative regime) the exponential of the dilaton with the string coupling parameter.

The above result is valid even if ϕ is not a constant, in which case $\exp \phi$ still plays the role of a *local* effective coupling. In general, in fact, the dilaton is a field that can take different values in different regions of space and at different times. The same will be true for g_s^2 and all the other coupling constants, i.e., for the "charges" that determine the strengths of the various forces, and which are obtained from g_s^2 in the context of string models of unified interactions.

Concerning this point, there is a nice analogy with the transition from Newton's theory of gravity to Einstein's. In the context of general relativity the rigid (Euclidean) geometry of Newtonian gravity becomes "soft", i.e., variable in space and time; the strength of the gravitational interaction remains "rigid", however. In a similar way the transition to string theory removes this residual rigidity because, thanks to the dilaton, even the gravitational strength (as well as the strength of the other forces) becomes "malleable" and space-time dependent. Note that it is just the variability of the dilaton (and the couplings) at early times that provides us with the new cosmological scenarios introduced in the last chapter.

Let us now go back to the physical effects associated with the finite size of strings. Given their tiny extension, it is evident that fundamental strings cannot be distinguished from point-like objects in any process where the typical length scale is much greater than L_s. On the other hand, current experiments involving particle accelerators are unable to resolve distances much smaller than about 10^{-15} cm. This means that they are only sensitive to length scales much greater than L_s, if we assume for the string length the standard value of about 10^{-32} cm suggested by unification models. Hence, the observed processes can be suitably described by applying the formalism of quantum field theory following the standard model of elementary particle physics, without any reference to string theory.

The same conclusion applies to the gravitational force if we limit ourselves to sufficiently flat portions of space-time, i.e., space-time regions whose curvature radius is much bigger than L_s. According to the standard cosmological model, however, going backwards in time towards the Big Bang, the curvature radius (i.e., the inverse of the curvature shown in Fig. 1.1) can reach extremely small values. In particular, taking the ratio between L_s and the speed of light c, it is possible to estimate the time at which the curvature radius of our Universe coincides with the string length. The result is about 10^{-42} seconds after the Big Bang (a time ten times bigger than the elementary Planck time L_P/c).

This is the time *after which* we may trust the cosmological predictions of general relativity. Before such times, the curvature radius of the standard model was indeed smaller than L_s. Hence, the extensions of strings (and of all particles) were greater than the curvature radius and could not be neglected by any means. This implies that, in this regime, the geometry of the Universe should be described by adding those quantum corrections and those extra degrees of freedom predicted by string theory to the equations of general relativity.

The origin and the possible form of those corrections will be discussed at the end of this chapter. The point we wish to stress here is that, as a physical consequence of such corrections, we may expect that the curvature radius cannot become smaller than L_s, i.e., that the curvature scale cannot exceed a maximum value controlled by the reciprocal of the string length parameter, $1/L_s$, therefore avoiding a possible singularity. This expectation is linked to an extremely important property of quantum strings, the already mentioned duality symmetry that was introduced and applied in the last chapter, and will be discussed in more detail here.

To begin with, let us consider a point-like object which is constrained to move along a circle of radius R, following the laws of classical mechanics. We can say that the point somehow feels the dimension of the circle. To complete a round trip, for instance, it takes more time on a large circle than on a small one. The sensitivity to the size of the circle remains valid in the framework of quantum mechanics, despite the fact that the quantization of the moving point-like object forces the velocity (or, more precisely, the momentum of the particle) to take discrete values proportional to the reciprocal radius $1/R$.

Consider now a string moving on a circle. As far as classical motion is concerned, the conclusion about the sensitivity to the size of the circle is similar to the previous case, but with an important difference due to the possibility of different types of motion. In particular, a closed string moving on a circle (or in a multidimensional generalization of it) can rotate (like a point), oscillate, and also wrap around the circle (you may think, for instance, of a cotton thread wrapped many times around a reel). The energy associated with wrapping, usually called winding energy, is proportional to the radius R and to the number of times the string is wrapped around (see Fig. 4.3). Hence, one could argue that strings feel the size of the circle in various ways.

However, this picture is drastically changed when the string motion is quantized. In fact, the total energy of the string must be computed by adding up the winding energy – which is an integer multiple of the radius – and the kinetic energy due to the rotational velocity – which, once quantized, is an integer multiple of the reciprocal of the radius (as in the case of a point-like object). An observer identified with the string would be confused, at this point. Since the kinetic energy and the winding energy cannot be distinguished by any possible means, by looking at the energy levels of the string he would not be able to establish whether the string is moving around a circle of radius R or one of radius L_s^2/R! (The fundamental length L_s must be introduced in order to guarantee the correct dimensions of length for both radii. See also footnote 1 of Chap. 3.)

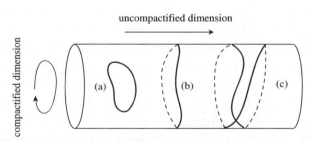

FIGURE 4.3 A simple example of a two-dimensional manifold in which the vertical direction is compactified to a circle, while the horizontal direction is not. A closed string can wind around the compact direction an arbitrary number of times, as illustrated in the figure. We have shown, in particular, three cases. (a) An unwound string, with winding number $m = 0$. (b) A string wrapped once around the circle, with winding number $m = 1$. (c) A string wrapped twice around the circle, with winding number $m = 2$

These two radii are totally indistinguishable for a string, in all respects. In other words, this means that it is possible to perform a transformation – called the dual transformation – which interchanges R and L_s^2/R without modifying any relevant aspect of string physics. This has an important consequence because, if R and L_s^2/R are equivalent, the effective minimum value of the radius is not zero but L_s, i.e., the value for which the two radii above coincide (this value is called the fixed point of the duality transformation). Indeed, when R is larger than L_s the effective radius experienced by the string coincides with R. When R is smaller than L_s, on the other hand, the effective radius coincides with L_s^2/R, which is still greater than L_s (see Fig. 4.4). This helps one to understand why L_s represents a sort of effective minimum length within the context of string theory.[2]

The above symmetry arguments (first developed by a group of Japanese physicists, Keiji Kikkawa, Masami Yamasaki, Norisuke Sakai, and Ikuo Senda in the 1980s) apply to circles, i.e., to the case of rigid geometries. However, as anticipated in the last chapter, the validity of the duality symmetry can also be extended in the presence of a time-dependent cosmological geometry (as pointed out by Arcady Tseytlin and Gabriele Veneziano), simply by replacing the radius of the circle with what we have called the spatial

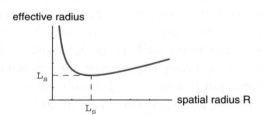

FIGURE 4.4 The effective radius given by the arithmetic mean between R and L_s^2/R, as a function of the "true" radius R of the circle. This plot shows that the effective radius felt by the string is always larger than the minimum length L_s, even when the radius of the circle around which the string is moving tends to zero

---

[2] These qualitative arguments are also supported by detailed computations concerning the gravitational deflection of highly energetic particles (with energies approaching the Planck scale), as shown by the analysis carried out at the end of the 1980s by Daniele Amati, Marcello Ciafaloni, and Gabriele Veneziano at CERN, and by David Gross and Paul Mende at Princeton University.

radius of the Universe (or, more technically, the so-called scale factor of the Robertson–Walker metric). Given the existence of a string-cosmology solution describing a model with spatial radius $R(t)$, duality then implies that there must be a dual cosmological solution with spatial radius equal to the reciprocal $1/R(t)$. Here, however, we encounter a crucial difference with respect to the case of a circle: the cosmological geometry varies in time, and satisfies a set of equations which (as we shall see below) are different from the ones predicted by general relativity. Those equations contain the dilaton and, when R and $1/R$ are interchanged, the dilaton must also be properly transformed in order to preserve the validity of the equations themselves. But we refer to Chapter 3 for a detailed discussion of this point and other properties of duality-related cosmological scenarios.

Here we want to comment on another important consequence of the finite size of the strings: the introduction of new energy states, i.e., of elementary new types of energy like the winding energy, whose possible existence is probably one of the most innovative features introduced by string theory into the physics of fundamental interactions.

The possibility for strings to wrap themselves around compact spatial dimensions, besides being at the heart of the duality symmetry, could indeed explain why, in a Universe which (according to unified theories) should have many spatial dimensions – at least nine, as we shall see in Chap. 10 – only three spatial dimensions have enormously expanded with respect to the string length, as is evident from our everyday experience. The mechanism leading to such an explanation has been suggested by work carried out by Robert Brandenberger and Cumrum Vafa at the end of the 1980s, in the so-called string gas cosmological scenario, later extended to the case of higher-dimensional extended objects (the brane gas scenario) by the joint contributions of Stephen Alexander, Robert Brandenberger, Damien Easson, Thorsten Battenfeld, Scott Watson, and others.

This basic idea stems from the fact that immediately after the Big Bang the Universe, in a multidimensional but highly compact configuration, should have been filled with a very dense gas of strings produced by the extremely high energies and temperatures. Not only were such strings moving at relativistic speeds, but they were also wrapped around all spatial dimensions, assumed to be compact. This network of wrapped strings (the so-called winding modes) prevented the Universe from expanding. Indeed, as soon as the expansion switched on (driven by the kinetic energy of the other strings) and the radius started to increase, the energy of the wrapped strings (which is proportional to R) soon increased; such strings then became dominant, balancing and overcoming with their tension the force sustaining the expansion, causing the geometry to contract back to its initial configuration.

The Universe was therefore in a multidimensional equilibrium configuration, with all dimensions equally extended but constrained to a compact size of the order of the string length L_s. So how was it possible for three out of nine spatial dimensions to pass through the net, as it were, and succeed in expanding without any constraint, leading to our currently observed Universe?

In order to answer this question we recall that at very high temperatures there should exist an equal percentage of strings wrapped in both orientations, i.e., wrapped strings and anti-wrapped strings (or winding and anti-winding modes), which annihilate each other by collisions, just as happens for matter and antimatter particles. Therefore, it may be that the wrapped strings gradually tend to disappear by colliding with their opposite counterparts, in such a way that, eventually, the "network" of winding modes breaks up, allowing the Universe to expand. If this is the case, however, why have only three dimensions been expanding?

The answer to this question is quite simple. In order to annihilate, strings must collide. If the space has too many dimensions it is likely that such collisions will never occur, even if the dimensions are compact. Let us think, for instance, of two point-like objects moving around a circle. Unless their velocities are exactly equal and have the same direction, the two objects are doomed to collide, sooner or later. If such point-like objects move instead on the two-dimensional surface of a sphere they may never met, even if their velocities are very different. In contrast, two one-dimensional objects like two strings have a finite probability of colliding even if they move on a sphere. And so on for higher-dimensional extended objects, in spaces with more and more dimensions. Iterating these arguments we arrive at the following general conclusion: given two p-dimensional objects (the so called p-branes that will be discussed in Chap. 10), the maximum number of compact spatial dimensions in which their collision becomes unavoidable is $2p + 1$.

For a point $p = 0$, we recover the result of our previous example relative to the one-dimensional circle. Strings, being one-dimensional objects, have $p = 1$. Hence two strings are very likely to collide and eventually annihilate each other within a space with at most $2 + 1 = 3$ dimensions (but not in higher-dimensional spaces!). The winding modes of wrapped strings were thus able to meet and completely annihilate only in a three-dimensional section of our Universe, and this is the reason why only three spatial dimensions have managed to escape from the string network, growing large and expanding to form our cosmos. Within the remaining six (or more) dimensions, on the other hand, wrapped strings have not experienced enough collisions, whence the network has not broken up, keeping such additional dimensions small and compact, confined to a distance scale of the order of the string length L_s.

At this point of the chapter, we can appropriately summarize the main results of our previous discussion by saying that it is just the finite extension of strings (compared with the point-like nature of classical particles) which provides the key for the new physical effects present in string theory.

Indeed, this finite extension allows the existence of new symmetries (e.g., duality) and new energy states (e.g., winding modes), which in turn suggests new cosmological scenarios and new mechanisms for dynamical determination of the effective dimensionality of our Universe. Furthermore, the presence of a minimal fundamental length L_s introduced by quantization should provide a way of avoiding the cosmological singularity at $t = 0$, and continuously joining our current standard regime to a primordial (pre-Big-Bang) inflationary regime. It is expected in fact that the quantization process, by providing strings with a finite extension, may also determine a maximum finite value for the curvature, therefore eliminating singularities in the quantum gravity regime – in just the same way as quantum mechanics has solved the singularity and stability problems of atomic orbits determining a minimum atomic radius, forcing the orbits to keep the electrons at a finite distance from the nuclei.

At this stage, however, a careful and expert reader could raise a question, by recalling that within the standard cosmological theory based upon the Einstein equations there are rigorous theorems (proved by George Ellis, Stephen Hawking, and Roger Penrose during the 1960s and the 1970s) stating that – under very general assumptions – it is impossible to avoid the initial singularity. If string cosmology can in fact avoid it, then string theory should yield gravitational equations that differ from those predicted by general relativity. What is the general form of these new equations, and how can they be derived from the theory?

Once again, the answer to these questions is deeply rooted in the symmetries of string theory. With regard to the first question, we recall that the duality symmetry associated with the inversion of the radius requires the presence of the dilaton field, which necessarily introduces a new scalar-type force into the gravitational equations. In the same way, generalized forms of duality are associated with the presence of other fields represented by antisymmetric tensors, which also contribute to the total gravitational force (see Chap. 10). In this context, the Riemannian metric of the curved space-time geometry is only one component of the total force coming into play. Not to mention the presence of fermionic components of the gravitational interaction, associated with the supersymmetry present in superstring models (see Chap. 10), making the resulting model of gravity an effective supergravity theory.

With regard to the second question, the answer calls into play another very important symmetry of string theory, known as conformal invariance (or Weyl invariance, or local scale invariance), which characterizes the motion of a string and its interactions, and which is absent in the case of a point-like object. We shall provide below a short illustration of the origin and properties of the conformal symmetry, but let us anticipate immediately that, thanks to this symmetry, the quantization of the string motion not only tells us what fundamental fields exist in nature (e.g., gravitational field, electromagnetic field, non-Abelian gauge fields, etc.) but also automatically gives us the *equations satisfied by these fields*. This is because the consistent quantization of an interacting string imposes rigid constraints on the fields interacting with the string.

This property probably represents the most revolutionary aspect of the theory with respect to conventional models based on the notion of elementary particle. In fact the motion of a point-like test body, even if quantized, does not impose any restriction on the external fields in which the body is embedded and with which it interacts. Such background fields can satisfy arbitrarily prescribed equations of motion, usually chosen on the grounds of phenomenological indications. We can think, for instance, of the Maxwell equations, constructed empirically from the laws of Gauss, Lenz, Faraday, and Ampere. It would be possible, in principle, to formulate sets of equations different from Maxwell's, but still preserving Lorentz covariance and other symmetry properties (such as the Abelian gauge symmetry) typical of the electromagnetic interactions. Such different equations might well be discarded, in the context of quantum field theory, but only for their disagreement with experimental results.

In the context of string theory, on the other hand, such alternative equations must be discarded *a priori*, as they would be inconsistent with the quantization of a charged string interacting with an external electromagnetic field. Indeed quantum string theory requires the electromagnetic field to satisfy a set of differential equations which, to lowest order, miraculously reduce precisely to the Maxwell equations (the same is true for the gravitational field, for non-Abelian Yang–Mills fields, and so on).

The above property of string theory is grounded in the geometrical properties of the two-dimensional surface spanned by the string evolving in the external space-time manifold – the worldsheet surface already mentioned, as illustrated in Fig. 4.1. Such a surface is curved, in general, and is thus characterized by a Riemannian metric associated with its intrinsic geometry. However, the area of this surface is an invariant, and does not change if the world-sheet metric is deformed by an arbitrary multiplicative factor which is local, i.e., variable in space and time.

This invariance is called conformal invariance, and represents a symmetry of the classical string motion. Thanks to this symmetry it is always possible to introduce a reference frame in which the world-sheet metric reduces to the flat Minkowski metric. Moreover, it is always possible to eliminate the "longitudinal" oscillations of the string, leaving only the degrees of freedom describing oscillations transverse to the string. Conformal invariance thus plays a crucial role in the process of determining the correct set of physical variables to be quantized, in order to obtain the correct quantum spectrum of physical string states.

When we have a test string interacting with any one of the fields present in its spectrum (for instance the dilaton field, or the gravitational field, or the electromagnetic field if the string is charged), we must then require, for consistency, that the conformal invariance (determining the string spectrum) be preserved by the given interaction, not only at the classical but also at the quantum level. In other words, the quantization of a string including its background interactions must avoid the presence of conformal anomalies, i.e., quantum violations of conformal invariance which is already associated with the world-sheet geometry at the classical level.

This observation leads us to the crucial point of our discussion: the only background-field configurations admissible in a string theory context are those *satisfying the conditions of conformal invariance*. Such conditions are represented by a set of differential equations corresponding, in every respect, to the equations of motion of the field we are considering. The field equations predicted by string theory – for any field, and in particular for the gravitational field – can thus be obtained directly by imposing conformal invariance on the quantum string interactions.

Unfortunately, however, such equations are hard to derive in closed and exact form for any given model of interacting strings. In practice, we have to follow a perturbative method: the quantized interaction of the string world-sheet with the background fields is approximated by a series of higher-order corrections as in standard quantum field theory, but with the difference that the fields are defined on a two-dimensional space-time, the string world-sheet.[3] The absence of conformal anomalies is then imposed at any order of this approximation, determining the corresponding differential conditions. As a result, the exact equations predicted by string theory for the background fields are approximated by an infinite series of differential equations, containing higher and higher derivative terms as we consider approximations of higher and higher order.

---

[3] Note that this approximation is not the same as the topological loop expansion illustrated in Fig. 4.2, since in this case the topology of the world-sheet surface is kept fixed.

To a first approximation (i.e., to lowest order) we then recover the second-order differential equations already well known for classical fields (i.e., the Maxwell, Einstein, and Yang–Mills equations, and also the Dirac equations for the fermion fields). To higher order, there are quantum corrections to these equations in the form of higher derivatives of the fields, appearing as an expansion in powers of the string length parameter L_s (also conventionally called the α' expansion, in terms of an equivalent parameter α' defined by $L_s^2 = 2\pi\alpha'$). Such corrections are a typical effect of the theory due to the finite extension of strings. Indeed, they disappear in the point-particle limit $L_s \to 0$, while they become important in the strong field limit in which the length scale of a given process (for instance, the space-time curvature scale in the case of gravity) becomes comparable with the string length L_s.

In conclusion, the new symmetries present in string theory tell us that the Einstein equations – and hence the gravitational equations to be used for the formulation of our cosmological models – are to be modified in two ways. Not only by the addition of new fields (like the dilaton), but also by the addition of quantum corrections due to strong fields (expansion in powers of L_s^2) and/or strong couplings (topological expansion in powers of g_s^2). In the context of pre-Big-Bang cosmology, both these corrections may play an important role in the transition to the phase of standard decelerated evolution, as we shall discuss in Chap. 8.

5. Inflation and the Birth of the Universe

The standard cosmological model, which describes the current Universe in terms of its matter and radiation components, and covers a seemingly long period of time – more than 10 billion years, from epochs preceding the synthesis of nuclear elements until now – legitimately represents one of the greatest achievements of twentieth century physics.

As already pointed out in Chap. 2, this model is based upon some crucial assumptions. One such assumption is that the geometry of the Universe and its time evolution are determined by the equations of general relativity. Another is that the whole particle content of our Universe can be described on cosmological distance scales in terms of a perfect fluid with two main components: matter and radiation, both uniformly distributed over space.

In this case the equations of general relativity successfully describe the expansion of this cosmic fluid which progressively cools down, according to the laws of classical thermodynamics, starting from an initial state characterized by arbitrarily high temperature and density. This expansion, slowed down by gravitational attraction, then "separates out" the various components of the cosmological fluid: heavy particles separate from the radiation, condense and form the matter structures (stars, galaxies) that we observe today, while radiation – which was initially the dominant component of the fluid – is diluted faster by the expansion, and tends to become subordinate to heavy particles and lumpy matter.

As outlined in Chap. 2, the standard cosmological model can rightly claim a number of important successes. First of all, this model explains important astronomical observations such as the redshift of galaxies described by the well-known Hubble law. In particular, it provides us with a consistent theoretical framework for computing the regression velocity of galaxies as a function of the distance, including also their possible deceleration or acceleration (depending on the equation of state of the dominant cosmological fluid), giving good agreement with astronomical data.

Moreover, the standard cosmological model can account for the primordial formation of light elements (the process of nucleosynthesis), since it provides a sufficiently hot environment for the required nuclear reactions to take place among the components of the primordial gas of particles. In addition, this model explains the existence of a cosmic background of electromagnetic radiation in thermal equilibrium, in full agreement with the progressively more precise measurements carried out today, confirming the presence of cosmic radiation with a black-body spectrum and a current temperature of about 3 degrees kelvin.

Despite the numerous successes, there are some kinematic issues within the standard model that remain unsolved, in addition to the already mentioned problem of the initial singularity. These kinematical problems concern the high degree of isotropy, homogeneity, and flatness characterizing the current Universe, and the large entropy associated with its background radiation. Why is the spatial geometry the Universe today so flat (i.e., so similar to the geometry of a three-dimensional Euclidean space)? Why – barring some irregularities due to localized matter clumps – is the background radiation so uniformly distributed over the whole observable space? And – since the standard cosmological evolution is adiabatic, i.e., entropy-conserving – what is the origin of the large entropy encoded in this radiation?

Leaving the last question aside for the moment (it will be considered at the end of this chapter and in the next), let us focus our discussion on the other points, starting with the question about the curvature. If we select a spatial portion of the current Universe, and we (indirectly) measure its curvature, we find that the maximum allowed value for such a curvature, according to the most recent observations,[1] is a few per cent of the total space-time curvature (which within the Einstein theory is represented by the square of the reciprocal Hubble radius, H^2/c^2). At first glance, the fact that the two curvatures, if not of the same order, are at least of

---

[1] See, e.g., D.N. Spergel et al. (WMAP Collaboration): Astrophys. J. Suppl. **170**, 377 (2007).

comparable magnitude, might seem a quite reasonable and acceptable result.

The problem arises because in an expanding Universe the curvature of its geometry varies with time. According to the standard
model, in particular, the curvature is currently decreasing so that
its value was higher in the past. However, tracing the solutions of
the standard cosmological model back in time, one finds that the
curvature of the three-dimensional spatial sections of the Universe
(henceforth referred to as the spatial curvature, for short), determined by the reciprocal of the spatial radius R, grows much more
slowly than the space-time curvature, determined by the Hubble
parameter H. Hence, even if we start from a present configuration
in which the values of the two curvatures are comparable, going
deeply backward in time we necessarily end up with a primordial
initial configuration where the spatial curvature is much smaller
than the space-time curvature (instead of being of the same order
of magnitude).

This is not by any means a natural initial condition. Indeed,
let us visualize – even if improperly – the cosmological space-time
as the two-dimensional surface of a sheet of paper. We would have
a sheet that, instead of being laid down onto a plane, would be
wrapped like a narrow cylinder with one dimension (time) highly
curved, and the other one (space) almost flat. Thus, we may naturally expect some previous phenomena to have occurred, determining such a highly asymmetric initial configuration through some
peculiar mechanism. Indeed, it would not be satisfactory to assume
that our Universe was born in that form solely because this configuration is the only initial condition able to produce the current
cosmological state. This amounts to abandoning any attempt at
scientific explanation.

In order to explain why, sometime in the past, our Universe
was in a geometric state characterized by a spatial curvature much
smaller than the space-time curvature, the standard cosmological model needs modifications at early epochs. The introduction of
such modifications has led to the formulation of the so-called inflationary models, first developed at the beginning of the 1980s, starting from ideas almost simultaneously proposed by astrophysicists
like Alexei Starobinski and Andrei Linde in the Soviet Union, and
Alan Guth, Paul Steinhardt, and Andreas Albrecht in the United

States. The term "inflationary" assigned to these models arises because, within the framework they propose, the spatial part of the Universe at some point "inflates", expanding at a very fast rate. This feature is an essential ingredient of such models in order to explain the relative decrease in the spatial curvature, and to achieve (at the end of inflation) a geometric state that can suitably represent the initial configuration for the subsequent standard evolution (continued until the present epoch).

The problem that we have just outlined is also called the flatness problem. Another issue pertaining to the standard cosmological model, and closely linked to the previous one, is the so-called horizon problem, which can be formulated as follows. The current Universe appears to be homogeneous and isotropic over the whole scale of the horizon, which is determined by the Hubble radius c/H (see Chap. 2). Going backward in time the space-time curvature increases, i.e., the Hubble parameter H increases, so that the radius c/H of the Hubble horizon shrinks. A problem then arises because, according to the standard model, the spatial radius R of the Universe shrinks with time more slowly than the Hubble radius.

Indeed, at a given epoch in the past, the radius of the spatial portion of the Universe that we are presently observing was much bigger than the radius of the Hubble horizon at the same epoch (see Fig. 5.1). This implies that in the past, according to the standard model, different portions of the currently observable Universe were included in different horizons, and thus were unable to interact and communicate with one another. If this is the case, why are the physical properties of the present Universe (e.g., the temperature of the radiation background) the same everywhere, as if all portions of space in the past were in causal contact, exchanging signals and interacting as if they were included within the same horizon?

The flatness and horizon problems (as well as other similar problems, all pertaining to the standard model) are strictly linked to each other. Both can be solved by assuming that the primordial Universe, before entering the phase of standard evolution (described in Chap. 2), has undergone a phase (called inflation) during which its spatial radius R expanded in accelerated fashion – hence faster than the horizon radius c/H which, during the standard phase, increases linearly with time, as illustrated in Fig. 5.1.

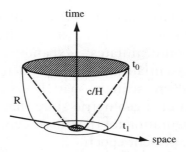

FIGURE 5.1 The *vertical axis* corresponds to the time direction, while the *horizontal axes* are associated with the spatial dimensions. At the present time $t = t_0$, the whole region of the Universe that we are able to observe is contained within the present Hubble horizon (*shaded region*). During the past (i.e., at any given time $t_1 < t_0$), the radius R of the current observable Universe was bigger than the corresponding Hubble radius c/H. The *dashed line* shows the time evolution of the horizon radius c/H, while the *solid curve* represents the qualitative evolution of the spatial radius R

The most conventional inflationary scenario is implemented by introducing a phase of de Sitter-like evolution, during which the spatial geometry of the Universe is subject to exponential expansion, while the space-time curvature (and thus the horizon radius) remain constant. The result is illustrated in Fig. 5.2, where the standard evolution is preceded by a phase of de Sitter inflation. With this modification, the currently observable portion of Universe at the beginning of inflation (i.e., at the time $t = t_2$) was all included within the same Hubble horizon. Therefore, all of its parts were initially in causal contact, being able to interact and give rise to a homogeneous and isotropic patch of space-time.

During the inflationary phase the spatial radius R expands faster than the Hubble radius c/H. We can say, using the common jargon, that the Universe goes "outside the horizon", i.e., it becomes larger than the horizon itself, while its degree of homogeneity and isotropy remain unaffected since – as already mentioned in Chap. 3 – outside the horizon all physical properties "freeze out". At the end of the inflationary phase (i.e., for $t = t_1$) this trend is inverted: R starts to increase more slowly than c/H, and the current observable Universe begins to "re-enter the horizon". The re-entry is eventually completed at $t = t_0$.

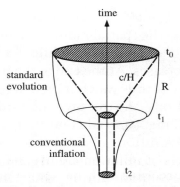

FIGURE 5.2 Qualitative sketch of the evolution of the spatial radius R (*solid curves*) and the horizon radius c/H (*dashed curves*) for a model of the Universe which includes a conventional de Sitter-type inflationary phase. The evolution from t_1 to t_0 describes the standard regime (exactly the same as in Fig. 5.1). The evolution from t_2 to t_1 describes instead the inflationary de Sitter regime, characterized by a constant Hubble parameter and an exponentially expanding spatial radius. The whole portion of the Universe that we are currently observing (the *shaded region* at $t = t_0$) was included within the same Hubble horizon at the beginning of inflation (the *shaded region* at $t = t_2$)

From Fig. 5.2 it is evident that a successful inflationary phase must last for a sufficiently long period of time. In fact, going backward in time, the time interval between t_1 and t_2 has to be sufficiently long for the whole currently observable Universe to have enough time to re-enter the horizon. The minimal required duration of the inflationary phase also depends on the size of the horizon (as can be seen from the figure), and thus on the curvature of the Universe at the onset of inflation. For instance, if inflation takes place close to the Planck scale (i.e., at the edge of the domain of validity of the standard model), then the required minimal duration is longer than in the case of inflation occurring at lower curvature scales, where H is smaller and the horizon radius is larger. However, if the amount of inflation is measured in units of the Hubble time $1/H$, then the same amount of inflation requires a shorter duration at higher curvatures, where H is larger and the inflationary process is faster.

It is interesting to check that the type of evolution described in Fig. 5.2 can also solve the flatness problem. To this end it will be enough to recall that the space-time curvature radius varies in time as the Hubble radius c/H, while the three-dimensional curvature radius varies as the spatial radius R. During inflation the space-time curvature remains constant, while the spatial curvature decreases as fast as R increases. At the end of the inflationary phase the space-time geometry is thus much more curved than its three-dimensional spatial sections, and this gives us a simple explanation for the origin of the "strange" initial condition characterizing the standard cosmological evolution.

The simplest models of inflation are usually characterized by a period of accelerated evolution during which the horizon either remains constant or slowly grows with time (this second case corresponds to the so-called slow-roll inflationary models). During the pre-Big-Bang phase typical of string cosmology models, however, the curvature increases, H increases (as we have seen in the previous chapters), and therefore the radius of the horizon tends to decrease. Even in this case, however, the spatial radius during the pre-Big-Bang phase undergoes an accelerated evolution which complies with the realization of an inflationary regime: the horizon exit of our portion of the Universe, in that case, is somehow even more rapid and efficient than in the conventional inflationary scenario (see Fig. 5.3). For this reason this type of inflation (first introduced in the 1980s by Deshdeep Sahdev, Eward Kolb, David Lindley, David Seckel, and others, quite independently from string theory and string cosmology models) is also called super-inflation.

In string cosmology models, the whole pre-Big-Bang phase may thus be considered as an inflationary phase, although an unconventional one, able to solve the kinematic problems of the standard cosmological model. The crucial difference – clearly illustrated by Figs. 5.2 and 5.3 – with respect to models of de Sitter-like inflation is that the initial size of the horizon at the beginning of inflation (i.e., the horizon at $t = t_2$) is much bigger for string cosmology models than for conventional (de Sitter-like) models. As the horizon radius is the reciprocal of the curvature, this feature of string cosmology models is a direct consequence of the fact that their initial curvature is very small with respect to the Planck scale. Indeed, in pre-Big-Bang models, the Universe starts evolving from an initial state quite close to the string perturbative vacuum (which is flat), unlike conventional models where inflation starts in a regime of very large curvature (and small Hubble radius).

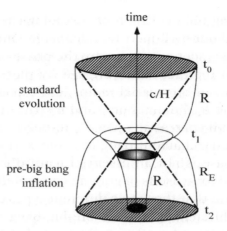

FIGURE 5.3 Qualitative sketch of the evolution of the spatial radius R (*solid curves*) and of the horizon radius c/H (*dashed curves*) for a model of the Universe in which the phase of standard cosmological evolution (from t_1 to t_0, the same evolution as depicted in Figs. 5.1 and 5.2) is complemented by a phase of pre-Big-Bang inflation (from t_2 to t_1), characterized by shrinking horizon and growing curvature. The figure illustrates the behavior of the pre-Big-Bang kinematics in both the string geometry, associated with an expanding spatial radius R, and the Einstein geometry, associated with a contracting spatial radius R_E. The two geometrical representations are physically equivalent, but in the Einstein geometry the initial horizon is larger, and inflation is longer (in general), than in the associated string geometry

The physical origin of such a difference is ultimately related to the dual symmetry present at the heart of string cosmology and to the fact that, in pre-Big-Bang models, inflation precedes (rather than follows) the Big Bang epoch. Note that, from a technical point of view, the prediction of a small value for the curvature at the onset of inflation should be considered as an advantage over models of inflation at large curvatures. In fact, if the curvatures are small, then the forces coming into play are weak, and the initial evolution is governed by well known low-energy physics and can be described in terms of simple, lowest-order equations. In the conventional inflationary scenario, on the other hand, the initial conditions are imposed at very high curvature scales, even inside the Planck scale, quantum gravity regime, where conventional low-energy results cannot be safely applied in general. This, in addition to the singularity problem, may also lead to the so-called trans-Planckian problem affecting the evolution of cosmological perturbations, as recently pointed out by Robert Brandenberger and Jerome Martin.

Finally, the reader may ask why the phase of pre-Big-Bang inflation has been represented in Fig. 5.3 by plotting two different types of behavior of the spatial radius, corresponding to the solid curves labeled by R (the inner one) and R_E (the outer one), describing expanding and contracting phases, respectively. (Note that the specular symmetry characterizing the contraction with respect to the standard phase is not an essential feature, but only a special choice suggested for reasons of graphic convenience.) It is thus important to explain that those two different types of behavior do not correspond to different models of pre-Big-Bang evolution; rather, they are two different (but physically equivalent) kinematic representations of the same model in terms of two different space-time metrics. Both representations are useful to provide an effective illustration of different aspects of the same scenario.

The expanding geometry (perhaps providing us with the most intuitive representation) uses as space-time metric the same metric "felt"[2] by a string present in the Universe during the pre-Big-Bang phase (the metric we also adopted at the end of the last chapter to discuss possible modifications of the Einstein equations introduced by string theory). This metric is also called the string metric, or string-frame metric. The contracting geometry (probably associated with the most conventional representation) uses instead the space-time metric felt by gravitons and dilatons (the fundamental particles of the theory), i.e., the same metric as would be used in the context of general relativity (it is in fact called the Einstein metric, or Einstein-frame metric).

It is always possible to switch from one representation to the other through a simple transformation which redefines the metric and other fields (performing, in particular, a local rescaling of such fields), without altering the physical phenomena, but simply describing them in terms of different variables. If we transform the geometry of a pre-Big-Bang model in this way we find, in particular, that the curvature keeps growing and the horizon radius keeps shrinking, but the expanding spatial radius R of the string metric

---

[2] Here "to feel" means to move freely (i.e., geodesically), following trajectories of extremum (minimum) length relative to the given geometry.

becomes a contracting spatial radius R_E in the Einstein metric, and vice versa. Thus, in the Einstein geometry the initial horizon is larger (and the duration of inflation is longer) than in the associated string geometry, as is clearly illustrated in Fig. 5.3. However, the two geometric descriptions are physically equivalent, and both provide a consistent description of the pre-Big-Bang scenario.

To complete the discussion of the main (practical and conceptual) differences between pre-Big-Bang inflation and conventional de Sitter inflation we should recall that a phase of conventional inflation cannot be extended arbitrarily far back in time (as already stressed in the first chapter). In such a case, working within the framework of general relativity (i.e., of classical gravitational physics), it is in practice impossible to answer questions about how inflation began, or what happened before inflation. Trying to answer such questions would necessarily require the methods of quantum cosmology (see Chap. 8); but even in that case the choice of the state of the Universe preceding the inflationary epoch is completely arbitrary, yielding the so-called boundary condition problem. Various proposals have been made for the initial state, with contradictory outcomes.

Within string cosmology, on the other hand, the possibility that inflation can effectively last for an infinite amount of time is not forbidden as in the case of de Sitter-type inflation. But in the case of a finite duration, the above-mentioned questions concerning the origin of inflation are well-posed, and can be answered entirely within the framework of string theory. The initial configuration of the Universe is then identified without ambiguity with a configuration approaching the so called string perturbative vacuum, free of interactions,[3] already introduced in Chap. 3. Detailed studies of the evolution of an initial, non-homogeneous perturbative state, carried out by Alessandra Buonanno, Thibault Damour, and Gabriele Veneziano, and followed by other studies by Alexander Feinstein, Kerstin Kunze, Miguel Vasquez-Mozo, and Valerio Bozza, have shed light on the possible mechanism that could ignite the inflationary phase.

---

[3] See Chap. 10 for a different assumption regarding the initial configuration, and in particular the initial values of the coupling constants, in the context of the so-called ekpyrotic scenario.

Indeed, as already discussed in Chap. 3, the initial state of pre-Big-Bang cosmology should be seen as a quite extended portion of space-time without any matter or forces, and hence extremely flat, empty, and cold. The further back in time we go, the weaker the interactions become, and the more the space-time geometry looks similar to the rigid geometry of special relativity. This does not mean, however, that the whole Universe is rigidly crystallized in a static configuration. Generally, small (classical and quantum) inhomogeneities may always be present, producing space-time fluctuations in both the metric and the dilaton field (as well as in all possible background fields in principle allowed by the theory).

To make an analogy we may think of the surface of a very quiet ocean, where nothing seems to happen. Only a few tiny waves propagate over the surface, occasionally colliding with other waves. If some of these collisions are strong enough, or if some wave becomes big enough to break up, some "foam" can be produced here and there, in a chaotic and random fashion. Similarly, in the primordial Universe, random fluctuations of the geometry (and of other background fields) could focus in a small spatial region a high enough energy density to trigger a local gravitational collapse, with a corresponding local "implosion" of both space-time and all forms of energy. A process of collapse similar to the one that, even today, could convert some dead stars into black holes,[4] i.e., "bottomless pits" of gravitational attraction where everything is swallowed up forever.

According to this representation of the initial cosmological state, our Universe could emerge from precisely this type of collapse, and thus correspond to the portion of the whole space contained within one of those black holes. Working with a simple but quantitative model one can then estimate that, in order to produce a Universe similar to the present one from the collapse, the initial size of the black hole must be at least of the order of the radius of an atomic nucleus (i.e., about 10^{-13} cm).

If we adopt the standard Einstein metric of general relativity, describing the process of collapse as a geometrical contraction, we

---

[4] For a description aimed at the general public of some properties of such highly collapsed objects see, for example, S. Hawking: *A Brief History of Time* (Bantam Books, 1988).

find that the above initial size of the collapsing region progressively shrinks, so that the resulting Universe becomes more and more compact. If instead we adopt the string metric, describing an expanding cosmological geometry, we find that the initial size, instead of shrinking to a point, grows in an accelerated inflationary fashion. The shift from contraction to expansion is due to the dilaton, which is the field responsible for rescaling the Einstein metric (i.e., for transforming from the Einstein to the string geometry), and which grows during the phase of pre-Big-Bang inflation, as already stressed in previous chapters. In both representations the curvature keeps growing, so that the curvature radius eventually reaches the minimum allowed value L_s marking the end of the phase of accelerated evolution. At that final stage, the initial size of the collapsing portion of space, described in terms of the string geometry, has increased from the initial 10^{-13} cm to something of the order of one tenth of a millimeter, corresponding exactly to the initial size required at the string curvature scale to reproduce our currently observed Universe after the period of standard evolution.

To help the reader to visualize this scenario, Fig. 5.4 shows a qualitative sketch of the model just described, where (excluding one of the three spatial dimensions) we provide three subsequent space-time diagrams of the collapse/inflation process. In that figure we can see how, at successive times (the various planes from bottom to top), the wavy sea representing the initial state has produced various types of gravitational collapses at various points, within spatial regions of different sizes. We can also see that, subsequently, one of those regions has inflated until it has reached a spatial size of 0.1 millimeters in correspondence of the string curvature scale, where quantum string effects are expected to stop inflationary expansion, converting it into the dual process (not represented in the figure), characterized by standard decelerated evolution.

According to this scenario, our Universe is included in one of those primordial "bubbles", and separated from other, possibly different Universes born from gravitational collapses characterized by different sets of initial parameters. This model for the birth of the Universe is reminiscent of some similar scenarios suggested from time to time by a number of theoretical physicists such as Valery Frolov, Moisei Markov, and Viatcheslav Mukhanov in the Soviet Union, and Robert Brandenberger, John Wheeler, and Lee

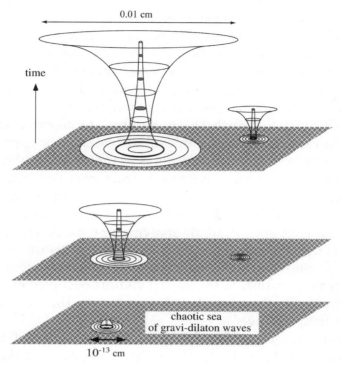

FIGURE 5.4 Inflation as collapse in pre-Big-Bang cosmology (the time co-ordinate increases up the axis). The various planes, from bottom to top, correspond to various sections of the space-time manifold filled with gravitational and dilaton fluctuations distorting the flat geometry of the perturbative vacuum. The space-time regions undergoing gravitational collapse are simultaneously represented using the string geometry (the outer expanding cones) and the Einstein geometry (the inner contracting cones)

Smolin in the United States. However, it differs substantially from the previous scenarios through the key role played by string theory and its dual symmetries.

It is interesting to note that, in this context, some of the properties of the current Universe can be directly traced back to the properties of the initial state giving rise to the subsequent inflationary evolution. In other words, encoded into the current observational data we can find the imprint of the Universe before the Planck era (i.e., of the cosmological state preceding the Big Bang and the quantum gravity epochs) – just as the final particles produced in a decay process contain (encoded in their quantum numbers) the imprint of the state of the system before the decay.

This possibility can be illustrated by focusing our attention on a key property of the current Universe: the entropy stored in the cosmic microwave background (CMB), which is measured – in appropriate units called natural units, where the speed of light c and the Planck constant $\hbar$ are both set to unity – by an extremely large dimensionless number, of the order of 10^{90}. In fact, the cosmic background of electromagnetic radiation is in thermal equilibrium (the temperature is the same everywhere, apart from tiny fluctuations that we can safely neglect for the purpose of this argument), and its energy thus follows the Planck statistical distribution (also known as the black-body spectrum). By applying statistical arguments one then finds that the entropy associated with this distribution is automatically nonvanishing. More precisely, the associated entropy is directly proportional to the spatial volume occupied by the radiation particles (in this case photons), and to the third power of the radiation temperature.

During the phase of standard (post-Big-Bang, post-inflationary) evolution, the volume increases as the third power of the spatial radius R (as the Universe is expanding), while the temperature decreases as the reciprocal of R (being redshifted like the frequency and the energy of the radiation). It turns out that the entropy is exactly conserved (i.e., the evolution is said to be adiabatic). Hence the standard cosmological model cannot explain the origin of the entropy currently associated with the CMB radiation. The value that we observe today was exactly the same value present at the beginning of the standard evolution.

An explanation of this entropy could possibly be provided by the inflationary dynamics and, indeed, the inflationary models were originally formulated to solve this entropy problem, along with the other problems already mentioned. To this end, the final stage of inflation is characterized by the occurrence of non-adiabatic processes that "heat up" the Universe, massively generating thermal radiation, and thus entropy, so as to agree with current observations. It should be noted, however, that the total value of the CMB entropy generated in this way, which could at a first glance appear to be a huge number when compared with macroscopic standards, is an extremely small quantity if compared with the entropy that could be associated with a Universe as large as the current one.

To understand this argument, pointed out by the famous mathematical physicist Roger Penrose in one of his books,[5] let us extrapolate the evolution of our Universe back in time according to the standard cosmological model, until we reach an epoch (that we shall call the Planck epoch) during which the radius of the Hubble horizon c/H, and hence the space-time curvature radius, were as small as the Planck length L_P (i.e., of the order of 10^{-33} cm). We already know, on the other hand, that going backward in time, the spatial radius R of the Universe decreases more slowly than the horizon radius (see Fig. 5.1). Using general relativity, we can estimate that, at the Planck epoch, the radius R was bigger than the horizon radius $c/H = L_P$ by a factor of about 10^{30}. Therefore, at that time the spatial volume of our currently observed Universe was filled with 10^{30} to the third power (i.e., 10^{90}) small, causally connected spheres of Planckian radius.

The currently observed entropy – measured by a number whose order of magnitude is just 10^{90} – can thus be reproduced by assigning one degree of freedom (i.e., one "bit" of information, to use the jargon of computer science) to any portion of the horizon area of Planckian size. This prescription is equivalent to providing every "small Hubble sphere" of radius c/H with the maximum entropy allowed by the so-called holographic principle, first conjectured by the theoretical physicist Gerard t'Hooft (Nobel Prize winner in 1999), and subsequently applied to cosmology by many others. It is also equivalent, as pointed out by Gabriele Veneziano, to assigning to each spatial volume enclosed within a Hubble horizon the same entropy as would be carried by a black hole of equal spatial extension.[6]

However, the above arguments yield an entropy of the order of 10^{90} only if they are applied to the CMB radiation during the Planck epoch. If we apply them during subsequent epochs, when the horizon radius was larger than Planckian, the corresponding value of the CMB entropy increases until it reaches today the extremely

---

[5] R. Penrose: *The Ermperor's New Mind* (Oxford University Press, Oxford, 1989).

[6] The black hole entropy, independently determined by Jacob Bekenstein and Stephen Hawking in the 1970s, is measured by the area of the black hole horizon expressed in units of the Planck length squared.

high value of 10^{122}, which is the value associated with the area of the present Hubble horizon (whose radius is equivalent to about 10^{61} Planck radii). Thus, we are led to the following question: Why is the currently observed CMB entropy – which seems to be a "maximal" entropy (according to the holographic principle, or to the black hole entropy) if its value is computed at the Planck epoch – so small when compared with the entropy that could be associated with the current horizon?

Now, it is well known from the basic principles of statistical mechanics that the entropy somehow measures the amount of disorder associated with a system. If the present entropy within our Universe is small, it means that our Universe behaves as a highly ordered system, i.e., a system that has not lost track of its origins, and which may still encode a lot of readable information about its past history. String cosmology, and pre-Big-Bang models in particular, seem to be able to provide a key to interpret the data about the CMB entropy in terms of the cosmological evolution preceding the Planck epoch.

Indeed, if the initial pre-Big-Bang phase is described as a contraction (in terms of the Einstein metric, where gravitons move along geodesics), the unavoidable outcome is a collapse and subsequent formation of a black hole. The horizon of such a black hole depends on the portion of the space that has collapsed, and the horizon area determines the maximum entropy associated with that portion of space (according to the recipe of Bekenstein and Hawking), which remains constant until the cosmological evolution is adiabatic.

The initial black hole horizon, on the other hand, coincides with the Hubble horizon which appears in the context of the string metric, where the pre-Big-Bang phase is described as an inflationary expansion. During the pre-Big-Bang phase, the radius of the Hubble horizon shrinks linearly (as shown in Fig. 5.3). Hence, if at the beginning of the process the whole entropy of the system is encoded into the surface of a single, large Hubble sphere (also corresponding to the black hole horizon), when the Planck scale is reached the total entropy – which always has the same value, since the process is adiabatic – is distributed among a large number of small Hubble spheres of Planckian radius, causally disconnected from one another.

The pre-Big-Bang scenario thus seems to be able to explain why one arrives at the Planck scale with an entropy which has the maximum value predicted by the holographic principle, applied to the total number of Hubble spheres contained in our Universe at that epoch. However, this entropy is of geometric type, i.e., it is associated with the horizon area and hence with geometric properties of the space-time under consideration. But why should the cosmic radiation, which is produced later and becomes dominant in the subsequent standard phase, be characterized by the same amount of entropy? String cosmology also seems to provide an answer to this question.

In the context of pre-Big-Bang models, the radiation that dominates the standard cosmological evolution is produced by amplification of the quantum fluctuations of the vacuum, according to a mechanism that will be described in detail in the next chapter. The amplification process already begins during the pre-Big-Bang phase, when the wavelength of the quantum oscillations becomes larger than the Hubble radius. As the horizon radius shrinks towards smaller and smaller values, oscillations with smaller and smaller wavelengths (hence higher and higher frequencies) get amplified, so that the energy and entropy of the quantum radiation included in a given portion of space grow progressively larger and larger.

At a given time, the amount of entropy of this quantum radiation thus depends on the reciprocal of the horizon radius. Detailed computation (performed mainly by two groups, one including Robert Brandenberger, Viatcheslav Mukhanov, and Tomasz Prokopec, and the other Massimo Giovannini and the present author) have shown that, within a given spatial volume, the entropy of this radiation is proportional to the number of Hubble spheres (of radius c/H) contained in that volume. On the other hand, the geometric entropy is proportional to the number of such spheres times their area in Planck units. The ratio between geometric entropy and radiation entropy, at any given time, is therefore approximately determined by the area of the Hubble horizon in Planck units.

This ratio, initially quite high, tends to decrease during the pre-Big-Bang phase, approaching unity when the space-time curvature gets close to the Planck scale. On the other hand – as we shall see again in Chap. 8 – we ought to expect the phase of standard cosmological evolution to begin when the contribution of the quantum corrections to the gravitational equations becomes significant, in particular when we must take into account the "back-reaction" of the radiation produced by amplifying the quantum fluctuations of the vacuum.

The relative weight of the quantum corrections is determined by the square of the space-time curvature in Planck units, i.e., by the ratio $H^2 L_P^2 / c^2$, which also expresses the reciprocal of the horizon area in Planck units, *hence the ratio between radiation entropy and geometric entropy*. Since the Universe tends to exit the accelerated pre-Big-Bang phase and become radiation-dominated just when the above ratio is of order one, it follows that the transition to the standard regime occurs precisely when the entropy stored in the produced radiation is equal to the geometric entropy, i.e., equal to the maximum entropy allowed by the holographic principle applied to the Hubble spheres.

If we accept the idea that the radiation corresponding to the currently observed cosmic background finds its primordial origin in the process of amplification of the quantum fluctuations of the vacuum, we can then explain why its entropy exactly saturates the maximum allowed value *when it is evaluated at the Planck epoch*. In such a context, the current value of the radiation entropy can be interpreted as the imprint left by the cosmological evolution during the epochs preceding the Big Bang (and the Planck era), and in particular by the size of the horizon of the initial geometric configuration from which our Universe has evolved.

To conclude this chapter we can say that the phase of accelerated evolution and increasing curvature, which could represent the primordial stage of our Universe according to pre-Big-Bang models, is a phase of inflationary type, able to overcome the shortcomings of the standard cosmological model, despite a type of kinematics which is profoundly different from the one characterizing other inflationary models of more conventional type. If we are convinced by the arguments presented in this chapter (and by others that will not be reproduced here, for the sake of simplicity) that such a phase could fully describe, in a complete and logically consistent fashion, the state of the Universe before the Big Bang – and could therefore constitute a physically acceptable model of primordial cosmological evolution – we are led to a question that may be regarded as crucial from a physicist's perspective: Are there phenomenological consequences, i.e., effects that are – at least in principle – observable, that could discriminate between pre-Big-Bang cosmology (or, more generally, string cosmology) and the more conventional inflationary cosmology? And are such differences observable today, in practice, given the current status of our technology?

The answer to these questions will be discussed in the following chapters.

6. The Cosmological Background of Gravitational Radiation

There exist a number of physical effects characterizing the various models of the primordial Universe, and allowing us to discriminate between the different scenarios. All these effects are linked more or less directly to the radiation production which characterizes the transition from the phase of accelerated, inflationary evolution to the decelerated phase typical of our current Universe.

During this transition, there is in fact a copious production of photons, gravitons, dilatons, and other kinds of particles – including the most exotic ones – which are peculiar to the various cosmological models. The number of particles produced as a function of their energy is called the spectral distribution or spectrum, and the fundamental feature of these spectra is the fact that they represent a sort of snapshot of the primordial Universe, taken at different angles (i.e., corresponding to different kinds of energy) and at different times. By combining the various snapshots (i.e., analyzing the information encoded into the various spectra) it is then possible to reconstruct, step by step, the past history of our Universe (unless the snapshots are of too low quality, i.e., the spectra are so weak that they escape detection).

In this chapter we focus attention on the production of gravitons, i.e., those particles carrying the gravitational force and representing the quanta of gravitational radiation (just as photons represent the quanta of electromagnetic radiation). Let us start by summarizing the basic properties of this type of radiation.

The possible existence of gravitational waves (which are absent in the context of Newton's theory) is likely to be one of the most interesting consequences of any relativistic theory of gravitation (hence, in particular, of general relativity). The production and propagation of gravitational waves is a phenomenon conceptually very similar to the one occurring in the electromagnetic context where, according to Maxwell's equations, the oscillations of electric and magnetic fields propagate from one point to another at the speed of light. Similarly, according to Einstein's equations, oscillations in the geometry can propagate from one point to another at a speed which (in vacuum) coincides with the speed of light.

Gravitational waves thus transmit information about how the gravitational field (i.e., the curvature of the space-time geometry) varies with time. Since the gravitational field is generated by masses and by their corresponding energies and momenta, it is the change in the status of motion of the gravitational sources – i.e., their acceleration – which generates perturbations of the local geometry, eventually propagating as a wave, and being transmitted to the whole surrounding space-time.

We can say, therefore, that the gravitational waves are produced by accelerated motion of masses, just as electromagnetic waves are produced by the accelerated motion of electrical charges. Furthermore, as in the electromagnetic case, there is no hypothetical medium (similar to the aether of pre-relativistic physics) which starts vibrating when a gravitational wave passes by. Gravitational waves can only be detected through the motion they induce in an appropriate system of test masses (just as electromagnetic waves are detected by the oscillations they induce in an ensemble of charges).

However, the analogy between gravitational and electromagnetic waves – apparently quite close – terminates here. Beyond the formal similarities mentioned above there are indeed various crucial differences, which will be stressed below.

A first important difference concerns the kind of acceleration required of a massive body, or a system of massive bodies, in order for them to emit gravitational radation. In contrast to the electromagnetic case, the distribution of masses and accelerations has to be sufficiently asymmetric. More precisely, the source of gravitational waves must have a non-zero quadrupole moment, which varies at a sufficiently fast rate (in particular, its third time derivative must be different from zero).

Under such conditions, for instance, a spherically symmetric cloud of gas, radially collapsing under the influence of the mutual attraction of the various molecules, does not emit any outward gravitational radiation. In fact, in spite of the fact that the single molecules are radially accelerated, the total quadrupole moment of the cloud turns out to be zero because of the spherical symmetry of the system.

Another crucial difference concerns the fact that it is impossible to block the passage of a gravitational wave (at least, using macroscopic shields made of ordinary materials). The reason is that the particles composing the shield start to vibrate under the influence of the impinging wave, in such a way that they exactly re-emit the wave absorbed by the shield. Hence, the gravitational waves keep propagating both within the shield and beyond it.

In the case of electromagnetic waves the situation is different, because of the existence in nature of charges of opposite sign. Thanks to its content of positive and negative charges it is possible for an appropriate shield to reflect an incident electromagnetic wave, by rendering null the oscillatory part of the field in a given region of space. However, gravitational masses of opposite sign seem to be absent from nature (at least, there has so far been no observational evidence for their existence). As a consequence, gravitational radiation cannot be shielded or reflected as simply as electromagnetic radiation.

A further difference concerns the tensorial character of gravitational waves, in contrast to electromagnetic waves, which are vectorial. This means that the intrinsic angular momentum carried by a gravitational wave is twice that carried by an electromagnetic wave of the same intensity.

In fact gravitons, i.e., the elementary particles associated with the quantum description of gravitational waves, have zero mass, like photons (indeed, they propagate in vacuum with the speed of light), but their intrinsic angular momentum is twice that of photons. This property, which emerges at the level of the quantum theory of gravitation, might seem to be irrelevant at the classical level. However, it is a crucial property of the gravitational interaction, since it is precisely through its tensorial character that the gravitational force between two masses of the same sign is attractive, rather than being repulsive as happens for the electromagnetic force between charges of the same sign.

We should add, as a final difference, that the intensity of the gravitational field is much weaker than the corresponding intensity associated with the electromagnetic field. Let us consider, for instance, the ratio between the static gravitational force and the electric Coulomb force mutually exerted between two protons, located at an arbitrary distance. This ratio is constant and is determined by the ratio between the square of the proton mass in Planck units – i.e., about $(10^{-19})^2$ – and the so-called electromagnetic fine structure constant α, whose value is about $1/137$. The result is a tiny number of order 10^{-36}. The fact that the gravitational force is so weak, together with the quadrupole-like nature of the corresponding radiation, imply that the emission of gravitational waves is a negligible process if compared with analogous processes associated with electromagnetic or nuclear forces. This weakness also explains why gravitational waves have not been directly observed so far in any laboratory.

In order to understand just how weak the power (i.e., the energy per unit time) emitted in the form of gravitational waves actually is, we may consider an oscillating mass (e.g., a pendulum, or a massive object attached to a spring, displaced from its equilibrium position). Its acceleration varies harmonically with time, so that its quadrupole moment and the time derivatives of the quadrupole moment are different from zero. Hence, the oscillating mass continuously emits gravitational waves. The emitted power (according to the theory of general relativity) is proportional to the mass squared, to the fourth power of the oscillation length, and to the sixth power of the oscillation frequency. However, the proportionality constant is the Newton constant divided by the fifth power of the speed of light! This is a truly tiny number, which gives gravitational waves of intensity too small to be detected by current instruments (at least if the mass and the frequency that we are considering are those typical of an oscillator produced artificially in a realistic laboratory).

This example suggests that in order to have more intense gravitational waves we should consider oscillations (or accelerations) of very big masses. Hence, we are naturally led to think that a class of promising sources of gravitational waves could be represented, in particular, by astrophysical processes in which the whole mass of a star is being accelerated. Indeed, a number of theoretical studies have shown that both high-velocity, close-orbit binary stars and collapsing/exploding stars (like the famous supernova observed in 1987) should radiate an intense flux of gravitational waves into the surrounding space.

Unfortunately, those sources are quite far away, and the radiation reaching us is so weak that it has not yet been directly observed by any of the currently operating gravitational antennas. However, there has been indirect evidence for the existence of those gravitational waves. In the binary system studied by Russel Hulse and Joseph Taylor, and associated with a pulsar (i.e., an extremely small and compact collapsed star), the orbits are shrinking with time. This happens because the system, emitting gravitational waves, loses energy, so that the stars tend to fall toward one another. Now, the corresponding decrease in the orbital radius which has been observed is in full agreement with the prediction of general relativity in the case of gravitational wave emission. Thanks to this discovery, the two astrophysicists were awarded the Nobel Prize in 1993.

Since the intensity of the gravitational radiation grows with the mass, the strongest source of gravitational waves we may ever envisage is certainly present at the cosmological level, and coincides with the Universe itself. Indeed, as already pointed out at the beginning of this chapter, the processes that take place during the primordial epochs – in particular, the more or less sudden deceleration driving the initial inflationary evolution to an evolution typical of the standard model – are associated with a copious production of gravitational waves which have filled the Universe, and which should still be present as relics of the "prehistoric" cosmological epochs.

For this kind of process, however, gravitational wave emission cannot be directly associated with the motion of accelerated masses – in fact, the primordial Universe may have been empty. Rather, it is the whole space-time itself which accelerates, and produces gravitational waves according to a mechanism called parametric amplification of vacuum fluctuations. Given the relevance of such a mechanism, which is quite effective not only for gravitational waves but also for other types of radiation (as we shall see in the next chapter), it is worth describing the basic principles in detail.

As already stressed in Chap. 2, the space-time geometry is fully determined, at the classical and macroscopic level, by the mass and energy distribution of the gravitational sources. However, at the microscopic level, there is still a tiny uncertainty in the local form of the geometry due to quantum mechanics, according to which all types of fields (including the gravitational field, and hence the geometry) can fluctuate. That is to say, they may undergo small local oscillations which, for a sufficiently short time interval, may drive them away from the value classically assigned to the field at a given point.

Such extremely fast fluctuations in the geometry are different at different space-time points, and their average value is zero. They can be treated as tiny "virtual" gravitational waves which are not freely propagating, being continuously emitted and immediately reabsorbed, locally, by space-time itself. Thus, according to quantum mechanics, space-time behaves as a sea which, even though it may be quiet and appear flat if seen from afar, shows a huge number of tiny "ripples", continuously changing in a stochastic fashion, when viewed more closely.

The quanta of gravitational waves, on the other hand, are the gravitons. These small quantum disturbances of the geometry can thus be seen as due to "virtual" gravitons which are continuously produced and then suddenly destroyed. To avoid violations of various conservation laws (for instance the conservation of momentum, which is always valid), these gravitons must be produced and destroyed in pairs. And here we arrive at the crucial point for the mechanism of parametric amplification.

If the geometry is static and without horizons (like the geometry of the flat and empty Minkowski space, for instance), the situation of the graviton pairs is stationary: pairs are formed and destroyed in a chaotic way, but on average the net result is null, i.e., the average number of gravitons is still zero.

However, if the geometry expands rapidly enough (as happens during the inflationary phase), it is possible for two gravitons, after being produced, to be "dragged" away from one another (thanks to the background expansion) so rapidly that they are no longer able to come back together and annihilate each other. A large number of gravitons become somehow uncoupled, and the net result is a copious production of gravitons (i.e., of gravitational waves) directly from the space-time itself.

This mechanism of gravitational wave production, which does not require the presence of accelerated sources, is based upon fundamental principles of quantum mechanics, and it applies not only to the gravitational field but to all types of fields (for instance, photons may be produced from the fluctuations of the electromagnetic field, and so on). Furthermore, the mechanism described here is related to the effect that produces quantum radiation from a black hole,[1] i.e., the radiation produced by a mass that has collapsed into a region of space so small that the gravitational force is strong enough to hold even light.

In fact, as explained by Stephen Hawking – the theoretical physicist who discovered that effect – even black hole radiation can be seen as due to the quantum fluctuations of the geometry, creating virtual pairs of particles in the region close to the horizon (i.e., close to the boundary of the region of space where light is trapped). Indeed, when one of the two particles is absorbed by the black hole, the other particle loses its "twin partner", which it would normally rejoin and annihilate, and can therefore move away from the point of its quantum creation.

The resulting effect is a flux of radiation flowing out to infinity, which seems to emerge just from the black hole horizon. The process of particle production occurring in the context of inflationary cosmology can be described in a similar way. The main difference is that, in the cosmological case, the two virtual particles are separated not by the black hole horizon but by the Hubble horizon associated with the phase of accelerated expansion.

A key feature of the resulting radiation is its spectrum, describing the radiation intensity as a function of its energy. As already pointed out at the beginning of this chapter, this quantity provides direct information about the state of the Universe at the epoch during which the radiation was produced. The spectral distribution can also be given in terms of the frequency rather than the energy, since the energy of a particle is proportional to the frequency of the associated (quantum mechanical) wave, with a proportionality constant given by the well-known Planck constant.

There are various ways of determining the radiation intensity as a function of the frequency (i.e., the spectrum). One of them is to compute the number of particles produced within each frequency interval, and then multiply this number by the particle energy. Another method, more intuitive and more suitable for illustration in

---

[1] See for instance S. Hawking, *op. cit.* in Chap. 5, footnote 4.

the context of this book, is to represent the emission of gravitational radiation as a result of the amplification of the microscopic fluctuations that spontaneously emerge everywhere, as a consequence of quantum effects. The intensity of this amplification as a function of the frequency immediately provides us with the desired spectrum.

Let us focus on the production of gravitational waves, i.e., on the quantum fluctuations of the geometry. Such fluctuations can always be decomposed into waves oscillating at different frequencies. However, given their quantum origin, the fluctuations satisfy a crucial normalization condition: the initial amplitude of these waves is proportional to their frequency, and hence inversely proportional to their wavelength λ.

In an expanding Universe all frequencies decrease (see Chap. 2), and the amplitude of those oscillations therefore decreases with time, while λ increases. On the other hand, the radius of the Hubble horizon remains constant during a phase of standard inflationary evolution (it can also increase, but more slowly than λ); alternatively, it can decrease, as happens in some string cosmology models (see Chap. 5). In any case, even if the oscillations of the geometry initially have wavelengths much shorter than the Hubble radius c/H, it is inevitable that the two length scales will eventually become equal.

From that time it is no longer legitimate to talk about oscillations, given that λ is greater than the horizon. The oscillation of this wave turns out to be invisible to all physical effects and to all causally connected observers. Hence, the amplitude of the wave remains "frozen" (as happens to strings in the scenario described at the end of Chap. 3) at the value it had when $\lambda = c/H$. On the other hand, the amplitude of the oscillations is inversely proportional to λ, so the final amplitude is proportional to the value of H (i.e., the curvature) at the time of freezing.

The waves "de-freeze" after the end of inflation, when the standard phase begins and H starts decreasing. Then oscillations take place once again, and their amplitude starts to decrease again. However, the freezing has prevented the amplitude from decreasing for a certain time interval, thus producing an effective amplification of the wave. The intensity of the final wave, which determines the amplification in the various frequency bands (and thus the spectrum), depends upon the value of the wave amplitude at the freezing time, which in turn is determined by the value of H during the accelerated phase.

It must be noted, at this point, that waves with different frequencies will be frozen at different times. In general, the higher the initial frequency, the smaller the wavelength, hence the longer the time required for λ to increase enough to eventually satisfy the condition $\lambda = c/H$. Therefore, the final amplitude is the same for all waves only if H remains constant during the whole inflationary phase. If H increases with time, on the other hand, high-frequency waves will freeze much later, and will have a greater amplification than low-frequency waves; the opposite occurs if H decreases with time.

This conclusion, obtained for the amplification of the geometric fluctuations (associated with the process of graviton production), tends to be valid also for other types of fields. In general, we may summarize the previous discussion by saying that the frequency behavior of the spectrum tends to follow the time behavior of the curvature scale during the inflationary phase.

Therefore, within conventional inflationary models (where the curvature is either constant or decreasing with time, see Chap. 5) the particles produced in the process of amplification of the quantum fluctuations will have a spectrum which is either flat or decreasing with their frequency (or energy). In the context of string cosmology models, where the inflationary phase preceding the big bang is characterized by an increasing curvature scale, the resulting particles will have a spectrum which tends to increase with frequency, as pointed out by Massimo Giovannini and the present author.

It should be stressed that this crucial difference between string cosmology and the standard inflationary scenario brings advantages as well as disadvantages. The advantages (of phenomenological type) are quite evident: the more effective production at high energies yields to the formation of a cosmic background of relic particles which is more intense in the high-frequency regime, i.e., just where direct observation is easier, in principle. Moreover, thanks to this more copious production, all the matter and radiation currently present in our Universe may be the direct outcome of the transition between the pre-Big-Bang and post-Big-Bang epochs, i.e., the direct result of the decay of the initial state (the string perturbative vacuum), as anticipated in the previous chapter.

Let us consider, in particular, the radiation which determines the geometry of the Universe just at the beginning of the standard cosmological evolution (i.e., soon after the big bang). In the standard cosmological model this radiation is introduced ad hoc; its presence is indeed one of the underlying hypotheses of that model. Within the conventional inflationary scenario the radiation is produced at the end of inflation as a consequence of quite complicated processes (phase transitions, inflaton decay, resonant oscillations, and so on), converting the potential energy of the dominant inflaton field into radiation. Within string cosmology, on the other hand, it is the energy of the initial perturbative vacuum that could transform itself into radiation; in that case, as discussed in the previous chapter, all the entropy we are currently observing could be obtained from quantum fluctuations of the initial state, amplified by the accelerated evolution of the space-time geometry.

A too copious production of particles may present some disadvantages, however. The resulting radiation may not satisfy present experimental constraints, and may even be inconsistent with the cosmological model itself. In fact, the energy of the resulting particles could be large enough to change the cosmological evolution, forbidding the onset of the standard regime.

Fortunately, the second issue concerning the consistency of the model may be intrinsically avoided within a cosmological framework based upon string theory. Indeed, the minimum fundamental length scale L_s defines a mass M_s which, using the system of natural units already introduced in Chap. 5, can be simply written as the reciprocal of the string length, i.e., $M_s = 1/L_s$. Since the reciprocal of a minimum represents a maximum, this mass (multiplied by the square of the speed of light) determines a maximum energy corresponding to about 10^{18} GeV, i.e., an energy one billion billion times bigger than the energy associated with the proton rest mass.

This energy is about ten times smaller than the energy corresponding to the Planck mass M_P, whose value (of order 10^{19} GeV) somehow represents the maximum energy which can be stored within the classical space-time geometry according to quantum mechanics. Beyond this energy scale the notion of space-time itself becomes uncertain. Within string cosmology, the maximum energy density of the produced particles is naturally determined by M_s. Hence, it cannot exceed the geometrical energy density – determined by M_P – and cannot have catastrophic consequences for the classical evolution of the space-time geometry on a cosmological scale.

Concerning the first issue, namely the compatibility of the resulting particles with the existing experimental constraints, the situation is more complicated, and it is worth addressing it in a detailed way. We shall focus the present discussion on a possible cosmic background of relic gravitational waves (or relic gravitons), which is the main subject of this chapter. No background of this type has yet been directly detected, but there are observations which already set constraints on its possible intensity.

For a clear discussion of such observations it is convenient to introduce the quantity Ω_G, which represents the energy density of the cosmic graviton background measured in units of critical energy density. The critical density is the total energy density associated with a model of the Universe whose three-dimensional spatial sections have a vanishing curvature. Since the currently observed spatial curvature is quite small, the critical energy density represents a realistic estimate of the full energy density of the current Universe.

There are three main kinds of observations which set direct or indirect constraints on the possible value of Ω_G. In all cases, the constraints emerge from the fact that a background of cosmic gravitons can be represented classically as an isotropic sea of gravitational waves distributed over all possible wavelengths, traveling and intersecting each other in all directions in a chaotic/stochastic fashion. These gravitational waves are perturbations of the space-time geometry, propagating at the speed of light, and universally coupled (even if very weakly) to all kinds of energy. Their presence thus produces a sort of cosmic disturbance, or background noise, on the homogeneous and isotropic large-scale geometry of the present Universe.

The most famous constraint on Ω_G probably comes from the high degree of isotropy of the cosmic background of electromagnetic radiation – the CMB "black body" radiation – which currently fills the Universe. The existence of a gravitational-wave background, distorting the geometry, would also induce a distortion of the isotropy of the CMB radiation. The anisotropy measured[2] in 1992 by the COBE (COsmic Background Explorer) satellite implies that, if there were gravitational waves in the cosmic background with a wavelength of the same order as the current value of the Hubble radius (i.e., of the radius of the observable Universe today), then their corresponding energy density could not exceed one ten billionth of the critical value. In other words, for those gravitational waves, $\Omega_G < 10^{-10}$. (We recall that the current value of the Hubble radius is about 10^{28} cm, and corresponds to the distance that could be covered in about 10 billion years, traveling at the speed of light.)

Another limit comes from the observation of the regularity in the beat of pulsating stars, the so-called pulsars. A background of gravitational waves, distorting the space-time geometry, would also produce a distortion of the observed pulsar timing. The absence of such an effect – confirmed in particular by Victoria Kaspi, Joseph Taylor, and M.F. Ryba in 1994 – implies that $\Omega_G < 10^{-8}$, i.e., that the energy density must be less than a hundred millionth of the critical value, for gravitational waves whose wavelength is about 10^8 cm (i.e., about one light-year).

Finally, an indirect (but important) limit comes from nucleosynthesis, i.e., from those primordial processes leading to the formation of atomic nuclei and hence to the synthesis of matter in its current form. These processes, which took place more than ten billion years ago, could not have happened undisturbed in the presence of too many gravitons in the background. Detailed computations give us a limit on the current energy density of such gravitons: it must be less than one hundred thousandth of the critical value, i.e., $\Omega_G < 10^{-5}$, for any wavelength. This limit, as well as the two previous limits, can be applied without distinction to any graviton background of primordial cosmological origin, quite irrespectively of the particular production mechanism.

Let us now recall that the gravitons obtained by amplifying the vacuum fluctuations within the conventional inflationary scenario would represent gravitational radiation coming from a constant or decreasing curvature phase, and would therefore be characterized by an energy spectrum Ω_G which is constant or decreasing with

---

[2] J. Smoot et al.: Astrophys. J. **396**, L1 (1992).

frequency, respectively. Within the pre-Big-Bang scenario of string cosmology, on the other hand, the same gravitons would represent gravitational radiation coming from a growing-curvature phase, and would have a spectral energy density that increases with frequency. Given the above-mentioned observational constraints, it follows that the graviton spectrum of highest allowed amplitude for the various inflationary models can be represented as in Fig. 6.1, where we have shown Ω_G as a function of the frequency ω (expressed in hertz).

The standard – flat or decreasing – spectra have been well known since the 1970s, thanks to the pioneering computations performed by the theoretical astrophysicists Leonid Grishchuk and Alexei Starobinski. Those spectra are shown in the bottom left part of Fig. 6.1, and it is evident that their maximum energy density is

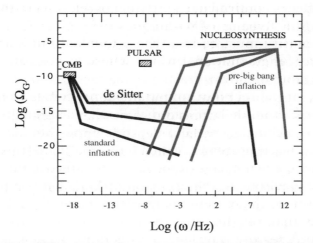

FIGURE 6.1 Possible spectra for a cosmic background of gravitons produced through the amplification of the vacuum fluctuations in the context of different inflationary scenarios. The plots show the logarithm of the energy density as a function of the logarithm of the frequency. The figure also shows the observational upper limits due to the CMB anisotropy, the pulsars, and nucleosynthesis. The spectrum labeled "de Sitter", and associated with a phase of constant-curvature inflation, marks the boundary between the decreasing spectra (standard inflation, *bottom left* part of the figure) and the growing spectra (pre-Big-Bang inflation, *top right* part of the figure). All spectra have been plotted at the maximum intensity compatible both with present experimental constraints and with theoretical predictions for string theoretical parameters

mainly constrained by the measured anisotropy of the CMB radiation. We note that for all these spectra, even in the limiting case of de Sitter inflation associated with a phase of constant curvature, there is a small jump at low frequencies, approximately located around the frequency scale 10^{-16} Hz, which is the proper frequency of a wave that begins to oscillate again just at the epoch of matter–radiation equality. This small jump is due to a further production of gravitons which takes place after inflation, and which is associated with the transition between the radiation-dominated phase and the matter-dominated phase (see, e.g., Chap. 2). As for the present discussion, that jump is not important. What is crucial is the fact that the de Sitter spectrum represents the maximum energy density distribution for a background of cosmological gravitons produced from the amplification of the vacuum fluctuations within conventional inflationary models.

The spectra obtained in the context of self-dual string cosmology models, and associated with a phase of pre-Big-Bang inflation, are shown in the top right part of the figure. Since they are increasing with frequency, they are not influenced by the low-frequency CMB constraint. In the low-frequency regime these spectra increase very fast, as the third power of the frequency, while at higher frequencies this increase flattens out somewhat, eventually reaching a peak value which essentially depends upon the ratio between the maximum string energy scale M_s and the Planck energy scale M_P (as shown in studies of Ram Brustein, Massimo Giovannini, Gabriele Veneziano, and the present author). Beyond that maximum value, the energy density of the background rapidly tends to zero (as shown in the figure).

The peak intensity of Ω_G for the pre-Big-Bang spectra is obtained from the fact that, at the end of inflation, the ratio between the graviton energy density Ω_G and the total radiation energy density Ω_{rad} is determined by the maximum curvature scale allowed by string theory, $H = M_s$, measured in Planck units and squared: $(M_s/M_P)^2$. The value of the string mass (or its reciprocal, the string length $L_s = 1/M_s$) is still uncertain. (So far, there have been no direct measurements of these quantities.) However, we do expect the value of M_s to be not greater than one tenth of the Planck mass, if string theory has to represent a unified model of all fundamental interactions (see the discussion in Chap. 4).

On the other hand, the total radiation energy density is currently about one ten thousandth of the critical value: $\Omega_{\mathrm{rad}} = 10^{-4}$. Given that the ratio between the graviton energy density and the total radiation energy density has not changed with time, it follows that the maximum energy density currently allowed for a graviton background produced by a phase of pre-Big-Bang inflation is about one billionth of the critical value: $\Omega_G = 10^{-6}$ (see Fig. 6.1). We observe that, despite the uncertainty in the value of the string-to-Planck mass ratio, the peak value of the pre-Big-Bang spectra is automatically compatible with the upper limit set by nucleosynthesis.

Once the peak value has been fixed, the spectrum of pre-Big-Bang gravitons shown in Fig. 6.1 is still characterized by two arbitrary parameters: the slope and the width of the high-frequency portion of the spectrum. In fact, the figure shows three possible examples of spectra, characterized by different slopes and widths. Such unknown parameters encode our lack of knowledge concerning the final stage of the pre-Big-Bang phase, where the curvature stops increasing and joins the subsequent phase of standard, curvature-decreasing evolution.

Actually, during those final stages, the curvature reaches values so high (close to the maximum allowed value) that it is mandatory to include purely "string-like" and quantum gravitational effects in the model. The low-energy approximation to the equations of motion is no longer legitimate: full, exact string-theoretical equations have to be implemented. This brings in technical difficulties that have so far prevented us from making detailed predictions, given the present status of knowledge about string theory. However, as this knowledge increases, it should be possible to develop a more complete model and to compute the high-frequency part of the spectrum explicitly, removing this kind of uncertainty.

It should be stressed at this point that the pre-Big-Bang spectra illustrated in Fig. 6.1 are those predicted by a particular class of "minimal models", where the height and the frequency position of the peak are both determined by the ratio $M_{\mathrm{s}}/M_{\mathrm{P}}$. Those models are characterized by three main cosmological phases. In the initial phase the curvature increases from the vacuum and the Universe "inflates" using the kinetic energy of the dilaton as a "pump" field. Then, during a second, intermediate phase – the so-called string phase – the Universe continues to expand in an accelerated fashion, while the curvature remains fixed around the maximum value allowed by string theory. Finally, the Universe enters the phase of standard evolution and decreasing curvature, where the resulting radiation fills the Universe and dominates over all other kinds of gravitational source (including the dilaton). Within this class of minimal models the peak of the spectrum is always fixed at a well-defined point in the plane of Fig. 6.1 (modulo a small uncertainty depending on our present ignorance about the exact value of M_s).

The frequency value (but not the height) of the peak could vary in the context of "non-minimal" models of pre-Big-Bang spectra where, after the string phase, the dilaton keeps growing even when the curvature is already decreasing, and where the resulting radiation becomes dominant only much later. In such models the peak of the spectrum, instead of being located at values of about one hundred billion hertz (as in Fig. 6.1), is set at smaller frequencies. This shift toward smaller frequencies would be experimentally desirable, since detection is more favorable at smaller frequencies. Indeed, as we shall see later, the peak sensitivity of currently operating gravitational wave detectors lies in a frequency range between ten and one thousand hertz. However, the non-minimal models appear to be less natural than the minimal ones. Barring possible "surprises", it seems unlikely that nature would have chosen this solution.

It is worth recalling that the height of the peak may also differ from (and, in particular, be lower than) the one shown in Fig. 6.1 – even in the context of minimal pre-Big-Bang models – if the gravitons, after being produced in the transition between the inflationary and the standard phase, were diluted by a so-called reheating phase, i.e., by a second, small Big Bang during which the Universe was heated up again by the generation of additional radiation and associated entropy. If we imagine the radiation as water, the Universe as a glass, and the gravitons as sugar, it is as if someone were to add more water to a glass already containing sugared water, and then to mix everything together: eventually the water turns out to be less sweet, i.e., the fraction of sugar has dropped. Similarly, the fraction of gravitons with respect to the full radiation content of the Universe would drop, and the present peak value of Ω_G would be lower.

This possibility does not seem to be very natural either, in a cosmological framework based upon string theory, where all relevant physical phenomena should take place at earlier epochs, near the string scale, rather than at lower energy scales. In any case, this effect would not dramatically influence the intensity of the graviton background. For instance, even if 99% of the radiation entropy that we are currently observing were due to any of those reheating processes subsequent to the end of pre-Big-Bang inflation, it can be shown that the maximum graviton energy density would drop from one millionth to one hundred millionth of the critical density, i.e., from $\Omega_G = 10^{-6}$ to $\Omega_G = 10^{-8}$. Such a value is still well above the maximum high-frequency value allowed for a phase of constant-curvature inflation (corresponding to the curve labeled "de Sitter" in Fig. 6.1), $\Omega_G = 10^{-14}$, i.e., one hundred thousand billionth of the critical density.

Finally, it should be noted that a growing spectral distribution is a rather typical – almost "universal" – prediction of string cosmology models alternative to standard inflation. However, a high-level intensity of the graviton background like that obtained in the context of the (minimal or non-minimal) pre-Big-Bang scenario is not a property of all string cosmology models. An important example of a low-intensity spectrum is given by the so-called ekpyrotic models, suggesting a string cosmology scenario alternative to the standard inflationary one, but different from the self-dual pre-Big-Bang scenario. The ekpyrotic scenario, which will be illustrated in Chap. 10, is indeed characterized by a growing graviton spectrum, but the associated peak intensity can reach at most the standard inflationary value $\Omega_G = 10^{-14}$, at a maximum frequency of about 10^8 Hz.

Given the various uncertainties characterizing the current theoretical models, and the lack of precise predictions pertaining to the exact form of the graviton spectrum that we would expect from a phase preceding the Big Bang, it seems appropriate to determine the so-called allowed region in the plane of Fig. 6.1, i.e., the region that undoubtedly includes the spectrum of background gravitons. This region corresponds to the portion of the plane "swept out" by the spectrum, varying all its parameters within the maximum allowed ranges (just as in the windscreen of a car, the region of glass swept by the wiper represents the allowed region within which the wiper itself has to be found).

In Fig. 6.2 we have compared the allowed region for a graviton background arising from the amplification of vacuum fluctuations within the self-dual pre-Big-Bang scenario (the region enclosed within the upper trapezium), with the allowed region of the standard inflation scenario (the region enclosed within the lower trapezium). It is easy to see that the first region is wider than the second by about eight orders of magnitude, since it allows a maximum spectral density of 10^{-6}, while in the de Sitter case the maximum, in the same frequency band, is only 10^{-14}. The figure focuses on the high frequency range, which is the phenomenologically relevant region for currently operating gravitational wave detectors; but the two allowed regions can be extended without modification down to

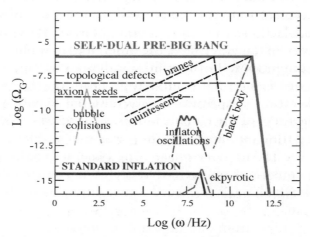

FIGURE 6.2 Allowed region for the cosmic background of relic gravitons produced through the amplification of vacuum fluctuations in the context of the pre-Big-Bang scenario (*upper trapezium*) and the standard inflationary scenario (*lower trapezium*). The figure also shows (with *dashed lines*) the graviton spectrum of the ekpyrotic scenario, as well as other possible types of spectra produced through mechanisms other than the amplification of vacuum fluctuations. Since these spectra are localized at high frequencies, they are not constrained by the experimental bounds on the CMB anisotropy

the millihertz range, so as to include the sensitivity band of other detectors, such as the space interferometers currently under study (see below).

It is instructive to observe that the large enhancement of the string cosmology region with respect to standard inflation can also be explained by recalling that the peak intensity depends upon the maximum curvature scale associated with a given model of inflation. In fact, within pre-Big-Bang models the spectrum increases with frequency, and the peak energy is fixed by the maximum attainable curvature. The latter can be expressed in dimensionless units by taking the square of the ratio between the string mass and the Planck mass. Given that the maximum of M_s is about one tenth of M_P (as previously discussed), we then obtain the maximum dimensionless ratio $(M_s/M_P) = 10^{-2}$.

Within conventional inflationary models, however, the spectrum decreases with frequency, and the peak energy is constrained by the low-frequency bounds imposed by observations of the CMB anisotropy. According to these observations there is a maximum value of the anisotropy that can be induced in the CMB temperature by the relic gravitons. The ratio of the possible temperature variation over distance scales of the order of the Hubble radius and the mean temperature of the radiation has to be (approximately) at most of order 10^{-5}. This anisotropy ratio, on the other hand, corresponds to the amplitude of gravitational waves with wavelength of the order of the Hubble radius, which in turn corresponds to the space-time curvature H/M_P (measured in Planck units) at the end of the inflationary phase. The observed CMB anisotropy thus sets a constraint on the curvature of conventional inflationary models, which cannot exceed a maximum value H such that $(H/M_P)^2 = 10^{-10}$.

Hence, going from pre-Big-Bang cosmology to the conventional one, we find that the maximum curvature changes from one hundredth to one ten billionth in Planck units, thus explaining the eight orders of magnitude of difference between the intensity of the spectra. It may be noted that, in conventional inflationary models, the scale associated with the maximum curvature is related to the grand unification mass scale M_{GUT} (rather than to the string mass M_s). The energy associated with this mass defines the energy scale at which the fundamental forces active at the microscopic level (i.e., nuclear, electromagnetic, and weak) are expected to be unified into a single force.

For the sake of completeness, Fig. 6.2 also shows the graviton spectrum produced in the context of the ekpyrotic scenario, which is approximately superimposed on the last part of the standard inflationary spectrum. Finally, we have plotted (with dashed lines) the spectra associated with other gravitational backgrounds, possibly obtained even within standard inflation, but produced by different mechanisms from the inflationary amplification of vacuum fluctuations. All these spectra have a negligible low-frequency tail, and so are not constrained by CMB and pulsar observations. Let us briefly illustrate these new types of possible spectra.

Amongst all those alternative backgrounds, a very intense one could be associated with the gravitational radiation produced by what are known as topological defects, i.e., geometric configurations characterized by particular symmetry properties: spherical symmetry for monopoles, cylindrical symmetry for cosmic strings, planar symmetry for membranes, and so on. Such objects could have formed during the phase transition characterizing the breaking of the grand-unification symmetry, i.e., the violent process that made the transition from a single type of force to the various components corresponding to all the fundamental forces that we now observe in nature. These objects may have survived this transition, and their vibrations could have produced a cosmic background of gravitational waves. In the cosmic string case – studied in particular by Alexander Vilenkin, Bruce Allen, Richard Battye, Robert Caldwell, and Paul Shellard – the resulting background is characterized by a spectrum which is flat at high frequencies, but much more intense than the de Sitter spectrum, as shown in the figure. We should mention that if the cosmic string network (generating the graviton background) has been produced in models of brane–antibrane inflation (see Chap. 10), then the peak intensity of the spectrum could be comparable to that predicted by pre-Big-Bang models. This possibility has been discussed recently by Edmund Copeland, Robert Myers, and Joseph Polchinski.

A globally flat graviton spectrum may also be produced in the context of the pre-Big-Bang scenario if the metric fluctuations, besides their primordial, direct amplification due to inflation, are indirectly amplified even during the post-Big-Bang era thanks to the presence of a cosmic background of particles called axions. These particles are also typical of string theory, and may be important for explaining the CMB anisotropy in a string cosmology context, as discussed in the next chapter. The spectrum produced by this "secondary" amplification has been computed by Filippo Vernizzi, Alessandro Melchiorri, and Ruth Durrer. It is flat at high frequencies, but less intense than the one produced by topological defects (see the curve labeled "axion seeds" in the figure).

Another possible background could be produced during the phase transition, typical of the so-called extended inflation models, signaling the end of the phase of accelerated evolution. Within these models, the end of inflation is characterized by the formation of "bubbles" in the space-time geometry (similar to the bubbles produced by shaking a bottle that contains a sparkling drink). These bubbles can collide and emit gravitational waves. This possible source of cosmic gravitational radiation was suggested by Michael Turner and Frank Wilczeck at the beginning of the 1990s. In this case, the intensity of the resulting background strictly depends upon the production temperature. In particular, the spectrum shown in the figure refers to a final temperature of about 10^9 GeV, i.e., more than a hundred billion billion degrees kelvin.

A graviton background could also emerge from the oscillation of the so-called inflaton, the field that ignites and maintains inflation within conventional models. At the end of inflation this fields enters an oscillating phase, and if the oscillations become resonant it is possible to produce a huge amount of radiation of any kind, including a gravitational component, as discussed by Bruce Bassett, Sergei Khlebnikov, and Igor Tkachev. This type of process is also called pre-heating of the Universe, referring to the fact the resulting radiation is not yet in thermal equilibrium, and that only at later stages will it be possible to define a cosmic temperature.

A further possibility is that, even within the standard inflationary scenario, metric fluctuations are amplified with a spectrum that is increasing at high frequencies (but not at low frequencies). This may happen for a class of models where the scalar field does not "freeze", i.e., does not lose all its dynamical properties at the end of inflation, but rather remains active, and can even significantly influence the expansion of the Universe by playing the role of the quintessential field (or dark energy) which seems to dominate current large-scale dynamics (see Chap. 9). We have shown in the figure (with the curve labeled "quintessence") the possible spectrum obtained in the context of four-dimensional models studied by James Peebles, Alexander Vilenkin, and Massimo Giovannini, and (with the curve labeled "branes") the possible spectrum obtained in the context of higher-dimensional models of braneworld inflation (see Chap. 10), studied by Varum Sahni, Mohammad Sami, and Tarun Souradeep.

Among the growing spectra shown in Fig. 6.2, we should also include the so called black body (or thermal) spectrum, with an effective temperature of about one degree kelvin, which could currently characterize a graviton background, possibly produced during the quantum gravity regime, when geometry and radiation were in thermal equilibrium. Within standard models, however, this background should have been significantly diluted by the subsequent inflationary expansion. Its current temperature should be much smaller than one degree kelvin, and its intensity should be reduced in consequence. Within the pre-Big-Bang scenario, however, a graviton black body spectrum like the one illustrated in the figure is not forbidden. On the other hand, it would not correspond to an epoch of thermal equilibrium, but to such a fast transition between the pre-Big-Bang and the post-Big-Bang phases that the curvature, once it reached the maximum at the string scale, would immediately start decreasing.

As is clearly shown in Fig. 6.2, all those possible additional backgrounds are generally stronger than the one obtained by amplifying the vacuum fluctuations in the context of the standard inflationary scenario. However, they are not stronger than the graviton background obtained in the context of the pre-Big-Bang scenario and all models characterized by equivalent kinematics.

In any case, given the plethora of possible spectra present in the frequency band shown in Fig. 6.2, one question comes to mind: Are any of these primordial backgrounds currently observable?

The straight answer is no. Current detector sensitivity is too low for this purpose. However, the current experimental limit (see below) is rather close to the theoretical value $\Omega_G = 10^{-6}$ which marks the boundary of the allowed region, i.e., the maximal expected background intensity. In addition, there are promising prospects for the not too distant future. In order to discuss this possibility, it is probably worth inserting a premise, albeit somewhat qualitative, providing a short description of the gravitational wave detectors and their fundamental operational principles.

The simplest detector we can think of consists of two masses mutually linked by a spring (at least two masses are needed, in order to make manifest the motion of one body relative to another). When a gravitational wave passes by, the two masses start to vibrate incoherently, i.e., their displacements from the equilibrium position are not simultaneously the same.

In fact, due to its quadrupole nature, the wave induces tidal forces in the two-mass system, so that the masses begin to move rhythmically to and fro, oscillating at the frequency of the incident wave. This effect induces oscillating tensions in the connecting spring. If these tensions can be amplified enough to be detected, we may be able to pinpoint the passage of the wave and measure its energy.

In practice, realistic gravitational detectors are not formed by two point-like masses, but by extended macroscopic bodies. The passage of the wave warps the space-time surrounding the body which works as a detector, so that the various constituent particles tend to follow the locally produced curvature. However, since different particles are located at different positions, each particle tends to follow different space-time trajectories. As a result, relative accelerations (also called geodesic deviations) are induced between the various points of the body, producing stresses which make the detector vibrate at the frequency of the incident wave.

Such a vibration tends to be damped by the friction present inside the body. However, a good detector is characterized by a characteristic frequency – its resonant frequency – at which the response to the incident wave is hugely amplified, and damping is ineffective.

With current technology the detectors (also called gravitational antennas), are mainly of two types: resonant bars and interferometers. Resonant bars are cylindrical metal objects (made, e.g., from aluminum), responding to passing gravitational waves with a vibration characterized by a typical resonant frequency of the order of one kilohertz (i.e., 10^3 Hz). The mechanical oscillations of the bar, induced by the gravitational wave, are transformed into electronic signals which are then efficiently amplified.

In order to observe the tiny vibrations of gravitational origin, it is mandatory to remove any other possible source of vibration, and in particular intrinsic, thermal oscillations. To this end, the bar is enclosed in an airtight container where a very high vacuum is created, and where the bar is cooled down to temperatures even smaller than one degree kelvin (this is the reason why such detectors are also called cryogenic detectors). We may well say that these detectors represent the coldest spot in the Universe, given that even deep intergalactic space turns out to be warmer! (As already observed, the black body radiation filling the whole Universe has a temperature of 2.7 degrees kelvin.)

By making use of all possible expedients, current technology would be able to detect oscillations with effective amplitude smaller than 10^{-16} cm, a length scale a thousand times smaller than the radius of an atomic nucleus. Nevertheless, no gravitational signal has yet been observed with absolute certainty by a resonant bar detector.

The first bar detector was studied and built by Joseph Weber at the University of Maryland in the 1960s. Today there are various bars operating at different locations around the globe, and the most powerful ones are in Italy: NAUTILUS, at the INFN (National Institute of Nuclear Physics) Frascati labs, and AURIGA at INFN Legnaro labs. These two detectors are the result of the evolution and refinement of a previous detector: EXPLORER, built and used by research groups of two universities in Rome (Roma 1 "La Sapienza" and Roma 2 "Tor Vergata"), but located at CERN labs in Geneva. EXPLORER is older and functions less well than the new versions, but it is still working. Other resonant antennas are ALLEGRO, at Louisiana State University, and NIOBE, at the University of Western Australia.

To get an idea of the main features of these detectors let us recall that they are cylindrical aluminum objects, with a typical weight of 2 300 kg, and typically three meters long. EXPLORER is cooled to a temperature of two degrees kelvin using liquid superfluid helium, while NAUTILUS and AURIGA may reach 0.1 and 0.2 degrees kelvin, respectively. Figure 6.3 shows a picture of the resonant bar AURIGA and associated experimental apparatus.

FIGURE 6.3 The AURIGA resonant bar, located at the Legnaro INFN labs, is enclosed within a large multi-layer container. In this photo the container is open, to give a full view of the cylinder located at the center. The role of the container is to provide the bar with thermal and seismic isolation. (Picture courtesy of the AURIGA collaboration)

The natural future evolution of bar detectors is represented by the spherical (or polyhedral) detectors, or resonant spheres (which are at present mainly in the design phase, however). The research activity for this type of detector began with the TIGA project at Lousiana University (the acronym stands for Truncated Icosahedron Gravitational Antenna), and continued with MiniGRAIL at Leiden University. Similarly to the bars, these spheres are made of metal and oscillate when a wave passes by. They can be filled or hollow, and in principle have various advantages over bars.

The first advantage is that one can determine the direction from which the wave originates, without comparison with another detector. Another advantage relies on the possibility of discriminating tensorial waves (associated with the propagation of gravitons) from scalar waves (associated with dilatons). This last property turns out to be quite interesting within string cosmology, which may also predict (as described in the following chapter) the possible formation of a cosmic background of dilatons, i.e., of scalar waves of gravitational intensity. The use of spherical antennas could therefore be appropriate, in particular, for hunting a possible dilatonic component of the background radiation.

Finally, hollow resonant detectors are expected to reach a sensitivity about two orders of magnitude better than can presently be reached by solid bar detectors. Particularly promising is the so-called DUAL detector, an Italian INFN project for a wide-band gravitational antenna consisting of a massive solid cylinder suspended inside a larger hollow one.

The other class of currently operating gravitational antennas are the interferometric detectors, where the masses that vibrate due to the passage of the wave are two large mirrors located at the endpoints of the arms of the interferometer. These mirrors reflect the light of a laser beam and the reflected light, properly combined, forms interference patterns that change when the passage of the wave induces oscillations of the mirrors.

The interferometer arms (along which the laser beam is traveling) are oriented at approximately 90 degrees to one another and are made of long metal pipes with a diameter of about one meter. Inside these pipes there is a vacuum, created to get rid of air molecules that would otherwise disturb the beam. However, unlike the bar detectors, the antennas are not cooled. Another important difference is that their maximum sensitivity is attained in a lower frequency band than the one for the bars (i.e., 10–100 Hz rather than one kilohertz).

There are currently three large interferometers already in operation. One of them is VIRGO, located in Italy (at Cascina, near Pisa), while the other two are part of the American project LIGO, located in Washington state (in the north-west of the United States) and in Lousiana (essentially at the opposite side, i.e., south-east). They are separated by 3 030 kilometers. There are also smaller interferometric detectors (with smaller sensitivity): GEO600, located in Germany near Hannover, and TAMA300, located in Japan (the numbers 600 and 300 refer to the length of their arms, measured in meters). Actually, the sensitivity of these instruments increases with the interferometer arm-length. The LIGO arms are both 4 km long, while the VIRGO arms are 3 km long. Figure 6.4 shows an aerial view of the VIRGO interferometer.

The main limiting factor for increasing the sensitivity of these antennas beyond a certain limit, at low frequencies, is the presence of seismic vibrations on our planet, representing a background noise which cannot be completely erased. To get round this problem, the joint project LISA between the European (ESA) and American (NASA) space agencies is currently under way. The idea of the project is to send an interferometer into space, following an orbit around the Sun.

FIGURE 6.4 The VIRGO interferometer located at Cascina, in the countryside near Pisa. The picture – taken from a plane – shows an aerial view of the current interferometric configuration. The two arms (3 km long, oriented at 90 degrees to one another) contain the vacuum pipes within which the laser beam is transmitted. (Picture courtesy of the VIRGO collaboration)

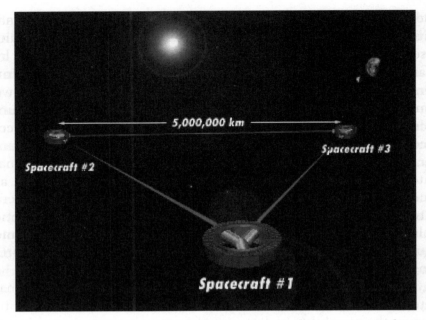

FIGURE 6.5 The LISA interferometer comprises three unmanned space-craft, separated by a distance of five million kilometers, and orbiting around the Sun. Each of the spacecraft emits and receives a laser beam from the other two. (Picture courtesy of the LISA collaboration)

The LISA interferometer comprises three unmanned space-craft located at the vertices of an equilateral triangle with side five million kilometers (see Fig. 6.5). Each spacecraft sends a laser beam to the others and receives one from the others. The fact that the arm length is so large, and that there are no seismic effects (there are no earthquakes in space!) should make it possible to reach extremely high sensitivities at low frequencies (i.e., around one thousandth of a hertz). Other space interferometer projects are DECIGO, proposed by a Japanese collaboration and operating in the frequency band from 0.1 to one hertz, and BBO, a constellation of four space interferometers (operating in the same frequency range as DECIGO), which is currently being investigated by NASA.

Finally, it should be noted that the overall frequency band covered by interferometers and resonant mass detectors (both in operation and projected) ranges approximately from the millihertz to the kilohertz. At higher frequencies this type of detector is useless, and an alternative possibility is to build gravitational antennas operating over the kilohertz to megahertz range using resonant electromagnetic cavities, as suggested in the 1970s by Francesco Pegoraro, Emilio Picasso, and Luigi Radicati at the University of Pisa. Work is in progress on the possibility of using two coupled microwave cavities, but the sensitivity presently attainable seems to be low compared with what is required to detect a cosmic graviton background. Other possibilities (currently under study) for the realization of very-high-frequency gravitational antennas are based on the use of electromagnetic waveguides, as recently suggested by Mike Cruise, or exploit the so-called Gertsenshtein effect – photon–graviton conversion in an external magnetic field – as proposed by Robert Baker, Zhengyun Fang, Fangyu Li, and Gary Stepenson, to detect waves with frequencies in the gigahertz range.

After this short review of the various types of gravitational detector currently in operation or still in the study phase, let us discuss the possibility of detecting the cosmological spectra shown in Fig. 6.2. To this end, it must be noted that a stochastically distributed ensemble of gravitational waves (as expected from cosmologically generated backgrounds) induces a stochastic signal in the detectors which is indistinguishable from other background noise that may already be present in the detector itself. Thus, for a reliable observation free from possible ambiguities of signal interpretation, at least two detectors are needed, in order to make a comparison and a cross-correlation of the registered data. A single detector can at most determine an upper limit for the energy density associated with a stochastic graviton background.

At present, the best level of sensitivity to a stochastic background of cosmic gravitons has been reached through the cross-correlated analysis of the data of the two LIGO interferometers.[3] It has been found that, for gravitational waves with frequencies in the range 51–150 Hz, there is no detectable signal associated with an energy density larger than about six hundred thousandths of the critical value. The cosmological graviton background must therefore satisfy the condition $\Omega_G < 6 \times 10^{-5}$ at frequencies around one hundred hertz. This is not a surprising result since, looking

---

[3] LIGO Scientific Collaboration: Astrophys. J. **659**, 918 (2007).

at Fig. 6.2, we immediately realize that the value $\Omega_G = 6 \times 10^{-5}$ is above the boundary of the allowed region. To have any chance of detecting a signal, the two cross-correlated detectors should be sensitive to at least the limiting value $\Omega_G = 10^{-6}$, which marks the boundary of the allowed region. This limiting sensitivity is expected to be reached by LIGO next year, clearly going beyond the nucleosynthesis bound at $\Omega_G = 10^{-5}$.

In order to penetrate well inside the allowed region, we must nevertheless wait for the realization of projects still under study, e.g., spherical resonant antennas, or hollow "dual" detectors, whose cross-correlation should allow us to push sensitivities to an energy density of about one ten millionth of the critical value. Another possibility is provided by the second, advanced generation of interferometers (e.g., Advanced LIGO), whose cross-correlation should eventually reach sensitivity to an energy density of about one ten billionth of the critical value. A third possibility is provided by space interferometers like LISA, which should achieve sensitivity to an energy density of about of one hundred billionth of the critical value, in a frequency band ranging between 3 and 10 millihertz.

For all these cases, the expected sensitivity is well within the allowed region. hence, future detectors could in principle detect a background of gravitons produced during a cosmological epoch preceding the Big Bang. A possible signal, extrapolated at the maximum (endpoint) frequency of the spectrum, could give us information about the peak value, thus providing the first experimental indication concerning the value of the fundamental string mass (about which there are, so far, only theoretical conjectures). But even the absence of signals in a given frequency band would provide information, since it would tell us either that the spectrum peaks at higher frequencies or that it is completely absent, thus imposing significant constraints on the parameter space of string cosmology models (and other models) of inflation.

The important conclusion of this chapter is that the background of relic gravitons, having such different intensities and properties for different models, provides a unique observational tool to test different cosmological scenarios. In particular, the background produced in the context of pre-Big-Bang models tends to be very intense in the high-frequency range, while it is so diminished in the low-frequency range that its contribution to the large-scale anisotropy of the CMB radiation (see next chapter) turns out to be completely negligible. On the other hand, the background produced within the standard inflationary scenario may contribute significantly to the CMB anisotropy, while at high frequencies – barring some peculiar mechanism of secondary graviton production, like those shown in Fig. 6.2 – is so low as to be barely detectable, even accepting the most optimistic predictions for the sensitivities of future detectors. Hence, a combined non-observation of gravitational wave contributions to the CMB anisotropy, together with a direct detection of relic gravitons at high frequency (in the allowed region of Fig. 6.2), could be interpreted as a strong signal in favor of the self-dual pre-Big-Bang scenario.

Needless to say, the direct observation of a cosmic background of gravitational waves would represent an event of importance comparable with the discovery of the cosmic background of electromagnetic radiation, first observed by Arno Penzias and Robert Wilson, who were awarded the Nobel Prize for this achievement. We may even argue that the discovery of the gravitational radiation would be more important. Indeed, while the photons represents the relics associated with the Big Bang explosion, these gravitons would represent much older relics which – according to string models – could even bring direct information about an epoch prior to the Big Bang.

7. Other Relics of the Primordial Universe

In addition to gravitational waves, there are other types of signal that could reach us from a primordial epoch preceding the Big Bang, and that could be (directly or indirectly) available to our present observation. Actually, as pointed out in the last chapter, the transition from an accelerated to a decelerated phase not only amplifies the fluctuations of the geometry (thus producing a gravitational-wave background), but also enhances the fluctuations of other fields. In this chapter we discuss in particular three important effects, all of them peculiar to string cosmology: the production from the vacuum of primordial magnetic fields, dilatons and axions. Let us start with the magnetic fields.

It is a well-known fact that all celestial bodies – the Earth, the Sun, the planets, up to galaxies and clusters of galaxies – have magnetic fields of various intensities. The origin of such cosmic magnetic fields is one of the issues that modern astrophysics has not yet completely resolved. We know, in particular, that magnetic fields can be produced by the rotation of electrical charges – as happens for instance in electric motors exploiting the dynamo effect. Since all celestial bodies rotate, their magnetic field could somehow be the result of a dynamo effect.

Even if this were true, however, the explanation would still be incomplete. In order to ignite the dynamo and trigger the mechanism producing the observed cosmic magnetic fields, the presence of a small initial field – also dubbed a magnetic seed – is essential. For the dynamos that we use in everyday life, (e.g., the small box with a rotating head producing the current for our bicycle light), the starting magnetic field is provided by a magnet. With regard to the dynamo that produces the magnetic field in a galaxy, the origin and the main features of the corresponding seed field are still largely speculative.

The simplest and most natural explanation is that the origin of the cosmic seed fields has to be traced to the vacuum fluctuations of the electromagnetic field, which are amplified during the accelerated evolution of our Universe (as first suggested by Michael Turner and Lawrence Widrow in 1988). Indeed, according to quantum mechanics (in particular, according to Heisenberg's uncertainty principle), the electromagnetic field in vacuum is never exactly zero but – as with all other fields – has some spontaneous, small-amplitude oscillations. An inflationary phase, during which the Universe expands in an accelerated fashion, could in principle amplify those oscillations and produce the magnetic seeds able to ignite the dynamo. After all, it is in exactly this way that we obtain the background of cosmic gravitons discussed in the previous chapter: the relic gravitational radiation is the result of the amplification of the vacuum oscillations of the gravitational field filling the Universe.

However, there is an additional issue for the magnetic fields. The Maxwell equations governing their evolution are characterized by a symmetry known as conformal symmetry. The presence of this symmetry implies that, for the evolution of the electromagnetic fluctuations, an isotropic, homogeneous and spatially flat space-time like the cosmological space-time is perfectly indistinguishable from a completely flat space-time. In other words, the electromagnetic vacuum oscillations are unable to "feel" the expansion of the Universe. Hence, they are not amplified, even if the expansion is accelerated, i.e., inflationary. The seed magnetic fields cannot therefore be produced in this way.

The above result refers to a classical cosmological framework based upon the Maxwell and Einstein equations. According to string theory, however, the electromagnetic field should also be coupled – in addition to gravity – to the dilaton, a crucial and unavoidable component of all string cosmology models. The vacuum oscillations of the electromagnetic field do not feel the expansion, but they are influenced by the dilaton evolution and can be efficiently amplified under the effect of a growing dilaton, as shown by Massimo Giovannini, Gabriele Veneziano, and the present author, and by the two French astrophysicists David and Martin Lemoine (brothers).

In fact, during the pre-Big-Bang phase the amplitude of the magnetic oscillations can increase, "dragged" somehow by the dilaton acceleration, until it reaches a "freezing" regime similar to the one described in the last chapter. During the subsequent phase of standard evolution, the fluctuations in the magnetic field turn out to be amplified with an increasing spectrum (as happens to gravitons), and begin to oscillate, re-entering the horizon. At re-entry, such fluctuations provide a homogeneous magnetic field within a causally-connected region of space. This field could act precisely as the seed for the dynamo that generates the galactic fields.

Explicit computations have shown that a pre-Big-Bang phase, as predicted by the minimal self-dual scenario, is able to produce primordial magnetic fields on cosmological scales with an amplitude more than appropriate for seeding the currently observed magnetic fields. There are also other, more or less contrived, explanations for the origin of the seed fields, but they require the ad hoc introduction of new fields and/or new types of couplings and/or new phases of cosmological evolution (the list of all models and all authors is too long to be reported here). The dilaton coupling, and its primordial accelerated growth, are instead natural predictions of a fundamental theory like string theory and its basic symmetries, i.e., duality.

This result may lead us to conclude that the seed production effect described above represents an undeniable success for string cosmology (and for the pre-Big-Bang scenario in particular). Pushing the argument forward as far as possible, we could even say that the existence of cosmic magnetic fields can be seen as an indirect proof for the occurrence of a cosmological phase dominated by the kinetic energy of the dilaton, and preceding the Big Bang. In this spirit, the analysis of the cosmological magnetic fields can provide indirect hints pertaining to the pre-Big-Bang phase, and can be used to test string models at the experimental level.

We should recall, in fact, that the coupling of the dilaton to the electromagnetic field may have different intensities, depending on the string model we are considering. As we shall see in Chap. 10, there are indeed five possible models of superstrings, with different physical properties. The pre-Big-Bang amplification of the electromagnetic and gravitational fluctuations for different string models has been compared by Stefano Nicotri and the present author. It has been found that for some models (for instance, type I superstrings) a large production of magnetic seeds is easily compatible with the production of a graviton background strong enough to be detected by near-future experiments. For other models (like heterotic superstrings), however, the two effects – large enough seeds and detectable gravitons – seem to be hardly compatible, at least in the context of the minimal pre-Big-Bang scenario. The cross-correlated study of magnetic and gravitational backgrounds – to be performed when gravitational antennas reach the required sensitivity – could therefore give us experimental information able in principle to discriminate between the various string models.

Let us now focus on another and quite exclusive feature of string cosmology, that has no counterpart in either the standard or the inflationary cosmology: dilaton production (studied by the present author soon after the formulation of the pre-Big-Bang scenario). In addition to gravitational waves, the transition from pre-Big-Bang to post-Big-Bang also amplifies dilatonic waves, i.e., the quantum oscillations of the dilaton field in vacuum. The outcome of such an amplification is the production of a cosmic background of neutral scalar particles (without charges, and with zero intrinsic angular momentum). Those particles, dubbed dilatons, should be characterized by a primordial spectral distribution very similar to the graviton distribution.

Actually, when computing the spectrum, one should take into account the fact that the dilaton fluctuations – unlike the gravitational fluctuations – are not freely oscillating, as they are coupled to both the scalar fluctuations of the geometry and the matter fluctuations. The corresponding equations are quite complicated and, up to now, have been solved only in some special cases. The outcome is a primordial spectrum, valid in the high-frequency regime, which increases exactly like the graviton spectrum, with a slope that is a model-dependent parameter. Such a parameter does not affect in any crucial way the total energy density of the dilatonic background, obtained by summing over all frequencies.

There is, however, an important difference with respect to the graviton case: the dilatons present in our Universe today could have a non-zero rest mass. If they have a mass, then the dilaton spectrum turns out to be modified at frequencies (i.e., energies) that are low with respect to the oscillation frequency associated with the rest mass of the dilaton. It is found, in particular, that this low-frequency part of the dilaton spectrum can be much more intense and flatter than the graviton (massless) spectrum. But why should dilatons be massive?

In order to answer this question let us recall that one of the milestones of Einstein's gravitational theory is the equality between inertial and gravitational mass. As a consequence of this equality all bodies should fall with the same acceleration (we may recall the mythical experiment carried out by Galileo Galilei at the leaning tower of Pisa). The fact that the motion of a test body within a gravitational field does not depend upon its mass – according to the so-called weak equivalence principle – has been checked with extreme precision, for both the terrestrial and the solar gravitational field. The various experiments have been carried out over distances ranging from the astronomical unit (corresponding to the radius of the Earth orbit, about one hundred million kilometers) down to the millimeter scale.

However, according to string theory, the gravitational interaction should always be accompanied by a second interaction carried by the dilaton field. The dilaton interaction may produce a force whose intensity should be of the same order as the gravitational force, at least to a first approximation. Hence, test bodies should feel both forces, and should be accelerated accordingly. However, while the response to the gravitational force is universal – since it depends on the ratio between the inertial and gravitational masses, which are the same for all bodies – the response to the dilatonic force depends upon the "dilatonic charge" of the body itself, which could be different for bodies characterized by different internal structures. For instance, aluminum and gold (or platinum) objects – like those used in high-precision tests of the equivalence principle carried out by Roll, Krotkov, and Dicke at Princeton in 1967, and by Braginskii and Panov at Moscow in 1971 – could fall with different accelerations due to the dilaton interaction.

On the other hand, an effect of this type has never been observed. Hence, string theory is consistent with present observations only if the effects of the dilatonic force somehow disappear at the macroscopic scales where the above-mentioned experiments have been carried out. In this respect, there are in principle two possibilities.

A first possibility is that the dilatonic force is short-range. If its effective range were to be smaller than the millimeter scale, in particular, such a force would not be observable in any of the experiments carried out up to now. On the other hand, the range of a force is inversely proportional to the mass of the carrier particle (in this case, the dilaton). If the particle is massless, the range is infinite: in order to have a finite range it is therefore necessary for the dilaton rest mass to differ from zero. In particular, to have a range smaller than the millimeter scale, the rest energy must be bigger than about 10^{-4} eV, i.e., the corresponding mass must be bigger than 10^{-37} grams (about one ten billionth of the electron mass).

A second possibility (suggested by the theoretical physicists Thibault Damour and Alexander Polyakov) is that the interaction between the dilaton and the macroscopic bodies is much weaker than the gravitational interaction. (This is not impossible, in principle, though more difficult to explain within string theory than the presence of the dilaton mass.) In this case the mass could be arbitrarily small and the dilatonic force, although long range, would not have been observed on a macroscopic scale simply because it would be too weak with respect to the current experimental sensitivities. Nevertheless, if the dilaton is sufficiently light, there could be interesting dilatonic effects on a cosmological scale (as we shall discuss in Chap. 9).

In this chapter we will mainly concentrate on the first scenario, where the dilaton coupling has a strength of gravitational intensity and, for the experimental consistency of string theory, the current value of the dilaton mass is sufficiently high to avoid detectable violations of the equivalence principle. Note that this last condition could be violated during the primordial evolution of the Universe (for instance, during the production of the cosmic dilaton background), since the dilaton could have acquired its mass only later at lower energy scales, through a symmetry-breaking process. Anyway, a non-zero value of the present dilaton mass modifies the present dilaton spectrum (rendering it different from the graviton spectrum), and yields two main consequences.

First of all, the dilaton speed tends to slow down after their production so that, if they are massive, all dilatons eventually become non-relativistic. In fact, even if the initial mass is quite small, as the Universe expands and cools down during the phase following the Big Bang, the kinetic energy of the resulting dilatons progressively decreases until it becomes smaller than their rest-mass energy. From that moment on the relic dilatons behave as a gas of almost static particles, with zero or negligible pressure, and their energy density (which becomes proportional to their mass) starts to increase with respect to the radiation energy density. It is then mandatory to impose constraints on the intensity of their spectrum (i.e., on the number and energy of the resulting dilatons), in order to prevent their energy from becoming so high that it would block the onset of the standard cosmological evolution, and the subsequent formation of the cosmological state that we now observe.

The second important consequence concerns the fact that, if dilatons are massive, they must decay, producing radiation (in particular photons, i.e., electromagnetic radiation). Such a radiation production would increase the entropy of the Universe and could affect nucleosynthesis (i.e., the production of nuclear matter), and even baryogenesis (i.e., the very early processes breaking the equilibrium between matter and antimatter, thus allowing the birth of the Universe in its present form, with a negligible amount of antimatter). The entropy produced through dilaton decay is inversely proportional to the square of the dilaton mass, and since such an entropy must be tamed – to allow baryogenesis and nucleosynthesis to occur as predicted by the standard model – it follows that the dilaton mass must be sufficiently high.

We thus arrive at a quite complicated, though interesting, scenario. On the one hand the dilaton mass ought to be sufficiently small to avoid the energy density of the non-relativistic dilatons exceeding a critical value, which would dominate the Universe, while on the other hand the dilaton mass must be sufficiently high to avoid violations of the equivalence principle, and to produce a sufficiently small entropy upon decay. This scenario is completed by the fact that the intensity of the dilaton background is not arbitrary, but depends (like the graviton background) upon the value of the string mass M_s.

The outcome is the existence of two bands of allowed values or, to use the jargon, two possible windows for the dilaton mass. These windows also depend upon the value of the dilaton charges, i.e., on the intensity of the coupling between dilatons and matter. In the case we are considering here – the case of couplings of gravitational intensity – the dilaton mass must be either greater than about 10 TeV (i.e., about 10^{-20} g, ten thousand times the hydrogen mass) or smaller than about 10 keV (i.e., about 10^{-29} g, one hundredth of the electron mass). In any case, there is the additional lower limit of 10^{-37} g, always valid, fixed by tests of the equivalence principle (see Fig. 7.1). The critical mass value of about one hundred MeV (i.e., 10^{-25} g), fixed by the process of dilaton decay, lies just between these two mass windows. Such a decay mass scale corresponds to a dilaton lifetime which is just of the same order of magnitude as the present Hubble time (which is the typical time-scale of our Universe in the present cosmological configuration).

The case of very massive dilatons (i.e., the mass window on the right of Fig. 7.1) is the least interesting one, from an observational point of view. In that case, in fact, all the dilatons of the resulting cosmic background would already have decayed (their lifetime is inversely proportional to the cubic power of their mass), and there would not be any (directly) observable dilatonic trace of the pre-Big-Bang phase. This seems to be the preferred case

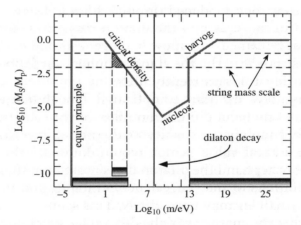

FIGURE 7.1 Phenomenological constraints on the dilaton mass, and allowed mass windows, for a matter–dilaton coupling of gravitational intensity. The constraints are plotted against the dilaton mass m and the string mass M_s on a logarithmic scale. The allowed region is below the constraints represented by *bold solid lines*, and within the *two horizontal dashed lines* delimiting the (theoretically expected) value of the string mass. The resulting allowed mass windows correspond to the *grey strips* placed at the bottom of the plot, along the horizontal axis. Finally, the *central vertical line* separates the mass values corresponding to a dilaton lifetime smaller than the Hubble age of the Universe (*to the right* of the line) from those mass values corresponding to a lifetime longer than the Hubble age (*to the left* of the line)

in the context of supersymmetric extensions of the fundamental gauge interactions, where the dilaton mass is directly linked to the mass of its supersymmetric partners (believed to be quite heavy).

If, on the contary, the dilaton mass lies in the left-hand mass window of Fig. 7.1, the resulting dilatons would still be "alive" today, and would represent a peculiar relic of the epochs preceding the Big Bang, with no counterpart in other kinds of cosmological scenario. In that case the energy density of the dilaton background could be very near to, or could even saturate, the critical density limit (see the small shaded triangle Fig. 7.1), thus representing a consistent fraction of the so-called dark matter density. The existence of dark matter (i.e., of matter that cannot be seen by ordinary optical telescopes and which fills the Universe on a cosmological scale) seems to be necessary to explain some discrepancies between the theory and the present astrophysical observations (as we shall discuss in Chap. 9). The detection of this kind of matter has been pursued for many years in various ways, but up to now without any decisive result.

This discussion of the possible cosmological effects of the dilaton unavoidably leads us to the following important question: If dilatons exist, how can they be detected? The answer depends on the strength of their coupling and on the value of their mass.

If their coupling to macroscopic matter has gravitational strength, and their mass is not too big, one exploitable effect could be the violation of the equivalence principle. As previously stressed, string theory does indeed predict that the various kinds of elementary particles may have different dilatonic charges. In contrast to the electric charge – which is universal, apart from the sign – the dilatonic charge of the proton, for instance, seems to be 40–50 times bigger than the dilatonic charge of the electron, as shown by studies carried out by Tom Taylor and Gabriele Veneziano in the 1980s. The total dilatonic charge per unit mass of a macroscopic body would then depend upon its internal structure (i.e., on the number density of protons and electrons). The possible fractional variations of the dilatonic charge between different bodies, due to their different atomic structures, are certainly small, of the order of one per thousand. However, if the range of the dilatonic force is not too much smaller than the millimeter scale, some experiments may hopefully observe such effects in the not so distant future.

Another possibility, in the same range of masses, is the experimental study of the mutual conversion between photons and dilatons in the presence of an intense magnetic field, in the laboratory. This possibility, however, does not seem to be within the reach of present technology, at least if the strength of the dilaton coupling is not much greater than the gravitational coupling.

We could also consider the use of the gravitational antennas illustrated in the last chapter. However, these can efficiently respond to dilatons only if the frequency associated with the dilaton mass is not higher than the frequencies of maximum sensitivity of the detector. Given the sensitivity range of current detectors (see Chap. 6), and observing that the frequency of one kilohertz corresponds to a mass of about 10^{-12} eV, we can easily conclude that such a detection method can be efficiently applied only in the case of ultra-light dilatons, coupled weakly enough to matter to satisfy present phenomenological constraints.

When the latter conditions are satisfied, there are in principle interesting prospects for the future detection of a cosmic background of massive dilatons. In fact, provided their spectrum is flat enough, they could induce an enhanced response to the non-relativistic part of the spectrum in both resonant mass and interferometric detectors, as shown by the studies of Carlo Ungarelli and the present author (with the collaboration of Eugenio Coccia for the spherical antennas). More precisely, the signal induced in two cross-correlated detectors could grow with the observation time much more rapidly than the signal produced by a massless background (like the graviton background). This effect could make a background of massive, non-relativistic dilatons detectable – in spite of the weakness of their coupling to matter – provided that the total energy density of the background is close enough to the critical value, and the dilaton mass lies within the sensitivity range of the antennas.

Given the various uncertainties, it seems wise to say that, at least for the time being, a direct experimental search for cosmic dilatons looks less promising than the corresponding search for gravitons, even though the history of physics has taught us that unexpected surprises are always possible.

The third effect we would like to discuss in this chapter is associated with the production of a cosmic background of neutral pseudo-scalar particles with interactions of gravitational strength, dubbed axions. In contrast to the case of graviton and dilaton production, such an effect could lead to already observed phenomena (as happens for the magnetic fields). The existence of axions is not peculiar to string theory. However, the axions produced during the pre-Big-Bang phase are characterized by a spectrum which – unlike that of gravitons, dilatons and photons – could be "flat" (or very weakly dependent on the frequency). As a consequence, axions could represent an indirect source of the observed anisotropy in the CMB radiation.

The cosmic microwave background (CMB) of electromagnetic radiation, frequently mentioned in this book, is a relic of high-temperature cosmological epochs, and its energy distribution follows a thermal (or black body) spectrum corresponding to a present temperature of almost three degrees kelvin. This background consists of a sea of photons (or electromagnetic waves) originating from the epoch when matter and radiation started to decouple, at a temperature of about three thousand degrees kelvin. As the temperature dropped below that value, the mean-free-path of photons became greater than the Hubble radius, so that the Universe became transparent to the electromagnetic radiation, and it has survived up to now undisturbed, providing us with a faithful imprint of that early epoch.

Those photons fill the whole currently observable space in an almost uniform fashion, apart from tiny variations of the local temperature whose fractional average value is of the order of one part in one hundred thousand. These temperature fluctuations make the background anisotropic, and are characterized by an angular distribution which is approximately flat (i.e., constant) at large angular scales (corresponding to distances of the order of the current Hubble radius), while it is oscillating at smaller angular scales.[1]

In the conventional inflationary scenario, the observed anisotropy of the CMB temperature is directly caused by the gravitational fluctuations of the geometry – in particular by their scalar part – amplified during the phase of accelerated expansion. The

---

[1] See, for instance, A. Balbi: *The Music of the Big Bang* (Springer, 2008).

oscillations of the gravitational field can in fact induce a small breaking of the spatial homogeneity and isotropy at the time of decoupling between matter and radiation, thus affecting the final temperature distribution. Such breaking would then propagate until today thanks to the so-called Sachs–Wolfe effect (from the names of the two astrophysicists who discovered it in the 1960s).

The above mechanism can successfully explain the observed anisotropies provided that the primordial gravitational fluctuations are amplified with a spectrum which is sufficiently close to being flat (also called the Harrison–Zeldovich spectrum), a requirement which is easily satisfied within the conventional inflationary scenario, as discussed in the last chapter. A phase of pre-Big-Bang inflation, on the other hand, amplifies the gravitational fluctuations (and also their scalar part) with a spectrum which tends to grow very rapidly with the frequency, and which cannot represent an efficient source of the observed CMB anisotropy. This is an issue for pre-Big-Bang models, since the anisotropy exists and must be explained.

A possible solution to this problem relies upon the fact that the anisotropy of the CMB radiation (i.e., the local fluctuations in its temperature) could be due *not* to the primordial geometric oscillations directly amplified by inflation, but rather to quantum oscillations of some other field.

In fact, if the fluctuations of such fields are amplified with a flat enough spectrum, acquire a mass, and subsequently decay (early enough), they can in turn generate a "secondary" background of geometric oscillations, characterized by the same spectral slope as the primordial background, and satisfying all the required properties to act as a source for the observed anisotropies. The subsequent process of anisotropy production is just the same as that of conventional inflation, the only difference being the secondary – rather than primordial – character of the geometric fluctuations. This mechanism, dubbed the curvaton mechanism (as it generates perturbations in the geometry, and then in the curvature properties of the space-time), has been proposed by independent and almost contemporaneous studies carried out by Kari Enqvist and Martin Sloth in Finland, David Lyth and David Wands in England, and Takeo Moroi and Tomo Takahashi in Japan.

But what field could play the role of the curvaton in a string cosmology context? Certainly neither the electromagnetic nor the dilaton field, since they are produced with a spectrum that increases too rapidly with frequency (like the spectrum of gravitational fluctuations). However, string theory contains other fundamental fields and, in particular, there is a field represented by an antisymmetric tensor whose fluctuations, the axions, could be amplified with the right (i.e., sufficiently flat) spectrum to source the observed anisotropy. The possibility of a flat axion spectrum was pointed out by the pioneering work of a group of theoretical physicists and astrophysicists including Edmund Copeland, Richard Easther, James Lidsey, and David Wands.

How can the axion spectrum be so different (flatter, in particular) from the other spectra so far analyzed? The answer relies upon the fact that the coupling of the dilaton to the axion field is exactly the reciprocal of the dilaton coupling to the gravitational fluctuations. Hence, whereas the accelerated expansion of the geometry tends to amplify the axion fluctuations, the accelerated growth of the dilaton tends to tame them, and the net result is a spectrum which increases much more slowly than the others. In particular, the final axion spectrum depends upon the number of expanding (or contracting) spatial dimensions, and if such expansion/contraction is isotropic then one finds that the spectrum is flat just for a certain "magic" number of dimensions equal to the so-called critical number of superstring theory (see Chap. 10).

At this stage, it is probably appropriate to point out that the number of spatial dimensions (three) that we normally consider (in both our everyday life and all physical experiments carried out so far) could not coincide with the number of spatial dimensions characterizing the Universe during the earliest remote epochs (in particular, during the phase preceding the Big Bang). This issue will be discussed in more detail in Chap. 10. For this chapter it will suffice to anticipate that the models trying to include all natural forces in a unified scheme (including gravity) agree on the fact that this unification, quite hard to achieve within a four-dimensional space-time, is actually easier to implement in spaces with a higher number of dimensions (superstring theories, for instance, require nine spatial dimensions, plus one temporal dimension).

So if the unified higher-dimensional theories are correct, how can we explain the fact that the Universe in which we live seems to have no more than three spatial dimensions? The standard answer is that an exact symmetry between all the natural forces, implemented in a multi-dimensional environment and at the high-energy scales typical of the primordial expanding Universe, cannot survive forever. At some time (i.e., below some energy scale) this symmetry breaks down, and three dimensions keep expanding while all the others (also called internal dimensions) "wrap around" themselves, becoming compact and so tiny that they are practically invisible in all lab experiments carried out so far. (This process is called spontaneous – or dynamical – dimensional reduction.) As a consequence, we live today in a Universe which effectively seems to have only three spatial dimensions, and where there are many types of force apparently quite different one from another (but see Chap. 10 for other possible explanations of why our space looks three-dimensional).

In the initial Universe, when the forces were united into a single root, all spatial dimensions were on the same level, and equally accessible. This legitimates the study of the primordial vacuum fluctuations in space-times with more than three spatial dimensions. One then finds that some spectra, like the graviton spectrum, are insensitive to the number of dimensions, while others, like the axion spectrum, depend strongly on the dimensionality of space, and may become flat in an appropriate number of isotropic dimensions.

Going back to the anisotropy of the CMB radiation we note that, if this anisotropy is indirectly generated by the decay of the cosmic axions produced by a phase of pre-Big-Bang evolution, then any anisotropy measurement carries direct information regarding the parameters and kinematics of that primordial phase.

We refer, in particular, to the already mentioned measurements of the COBE satellite, which fix the amplitude of the temperature fluctuations at large angular scales, and to the more recent measurements of the WMAP satellite, which determine the height of the first oscillation peaks, and the spectral slope of the geometric fluctuations generating the CMB anisotropy. These measurements may give us direct information about the amplitude and slope of the primordial axion background, which in turn depend, respectively, on the values of the string mass M_s and the number of dimensions (and their kinematics) of the accelerated epoch. We can then obtain important experimental constraints on models of pre-Big-Bang inflation, as discussed by Valerio Bozza, Massimo Giovannini, Gabriele Veneziano, and the present author.

For this purpose, it is important to stress that an axionic origin of the CMB anisotropy could be experimentally confirmed by its non-Gaussian statistical properties, arising if the primordial axion background decays early enough, before becoming the dominant cosmological source (as pointed out by David Lyth, Carlo Ungarelli, and David Wands). If, on the contrary, the mass is too small to forbid the decay, and the energy of the axion background always stay sub-dominant with respect to other gravitational sources, the axions can still contribute to the CMB anisotropy as seeds, i.e., as quadratic sources of the geometric fluctuations, as discussed by Ruth Durrer, Alessandro Melchiorri, and Filippo Vernizzi. In that case, however, the angular distribution of the temperature anisotropy turns out to be significantly different from that obtained from the standard inflationary scenario and from the curvaton mechanism. As a consequence (and according to present observations), any seed-like contribution is possibly allowed only as a subdominant component of the total observed anisotropy.

To conclude this chapter, it is interesting to observe that the phenomenological consequences of pre-Big-Bang models can be classified into three main classes, according to their observational chances. To paraphrase using well known science-fiction jargon (concerning encounters with extraterrestrials), we may divide the phenomenological consequences into first kind, second kind, and third kind. Effects of the first kind would be discovered by observations to be carried out in the next twenty to thirty years or so; effects of the second kind are relative to observations to be made in the near future (within a few years); finally, effects of the third kind are relative to observations already realized or currently being carried out.

An effect of the first kind, discussed in the last chapter, is the production of an intense background of relic gravitational radiation. Currently, gravitational antenna do not have enough sensitivity to detect such a background. However, the required sensitivity is expected to be reached in the not so distant future by detectors that are either being built or being planned.

As an effect of the second kind we may mention the total absence of contributions arising from the relic gravitational waves to the large scale anisotropy of the CMB radiation (because of the very steep slope of the graviton spectrum); but also, the presence of a small non-Gaussian feature in the spectral distribution of the anisotropy (because of its indirect axionic origin). There are satellites like PLANCK (see Fig. 7.2) which will be launched in the very near future, and will soon be able to perform high precision measurements of the fine-structure properties of the CMB anisotropy, on various angular scales. In this way we may expect to obtain

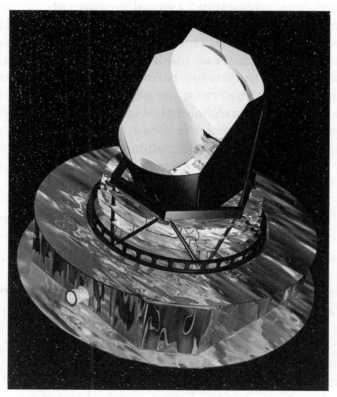

FIGURE 7.2 Artist's view of the PLANCK satellite (formerly called COBRAS/SAMBA), an ESA (European Space Agency) project for extremely accurate measurements of the anisotropy properties of the electromagnetic background radiation. (Picture courtesy of the Planck Science Team)

precise information about the two effects mentioned above, thus confirming or disproving the predictions of different inflationary scenarios.

An effect of the third kind could be represented by the production of magnetic seeds. The presence of cosmological magnetic fields requiring those seeds could be interpreted as an indirect confirmation of the string models able to produce them, in view of the difficulties in generating the magnetic seeds in the context of other inflationary models. Another example of an effect of the third kind could be represented by the CMB anisotropy measured by COBE and WMAP, under the hypothesis that its axionic (pre-Big-Bang) origin may be confirmed by future observations.

Finally, let us note that the phenomenology associated with the dilatons does not seem to fit any of the three above-mentioned kinds of effects, given the experimental difficulty involved in direct detection of such particles. However, there is one possible exception. The physical effects of the dilaton field may already have been observed (i.e., they could be of the third kind), if this field turned out to be the elusive quintessence dominating the present Universe and producing the observed large-scale acceleration (this possibility will be illustrated in Chap. 9). Up to now, recent and current studies do not seem to invalidate this interesting scenario.

8. Quantum Cosmology

The possible cosmological scenarios outlined in this book have been approached so far from a purely classical perspective, using concepts and ideas typical of the macroscopic world, like space, time, geometry, gravitational forces, and so on. The aim of this chapter is to introduce a possible alternative description of the cosmological evolution based on a quantum point of view, and using a framework where the Universe can be represented as a wave propagating in an abstract, multidimensional space dubbed superspace (no connection with the previously mentioned supersymmetry).

A detailed and rigorous explanation of this approach would clearly require some knowledge of quantum mechanics, which is not necessarily part of the scientific background of the typical reader, and whose introduction is beyond the scope of this book. Hence, our discussion will be grossly qualitative and approximate. Nevertheless, we hope to provide the reader with an appropriate overview of the methods and goals pertaining to the field of quantum cosmology.

Amongst the motivations suggesting the use of a quantum cosmology approach within string models of the Universe, we should mention first of all the difficulties we currently face when we try to give a quantitative and fully consistent description of the transition between the pre-Big-Bang and post-Big-Bang phases. Whereas the obstacles appear to be of a formal nature, they nevertheless have a physical origin. Indeed, they are rooted in the fact that the instability of the initial state (the string perturbative vacuum) yields to a phase (the pre-Big-Bang phase) in which the curvature and the strength of the gravitational force (and of all other forces) increase in an accelerated manner.

Hence, in order to make a transition to the standard cosmological phase where the Universe decelerates and becomes radiation-dominated, and where all natural forces are stabilized, we need a mechanism that curbs the initial increase of both the curvature and the dilaton. Otherwise, the Universe would necessarily reach a singular state with infinite curvature (like the one occurring in the standard scenario). Such a singularity would then completely detach the pre-Big-Bang phase from the current one. Without any physical connection, it would no longer make sense to relate the properties of the current Universe with those characterizing an epoch preceding the Big Bang. Furthermore, it would not be legitimate to hunt for possible experimental traces of such a primordial epoch in the cosmological backgrounds of relic radiation, as discussed in previous chapters.

The dynamics of a possible mechanism able to stop the increase of the curvature and to provide a transition from the pre- to the post-Big-Bang phase is quite complicated, as shown by all studies and computations so far performed. Actually, in addition to the above-mentioned effects, such a mechanism should be able to turn the kinetic energy associated with the geometry and the dilaton into thermal radiation. Furthermore, if the pre-Big-Bang phase is higher-dimensional, such a mechanism should be able to "freeze" the extra spatial dimensions, and possibly break the symmetries between the various forces.

According to string theory, those effects can hardly take place when the curvature is small and the couplings are weak. The transition seems to require a phase where the gravitational forces are so strong that the resulting particles are themselves able to modify the geometry, yielding what are known as back-reaction effects, introducing quantum corrections into the classical equations (the quantum loop corrections introduced in Chap. 4). Moreover, when the curvature is quite high, other corrections (the so-called α' corrections, see again Chap. 4) are induced by the fact that it is no longer legitimate to approximate string behavior by point-like objects. In addition, the dilaton may start to develop a strong self-interaction, generating a large potential energy density.

Taking into account all these effects, the full equations of string cosmology become so complicated that – up to now – it has been impossible not only to find their exact solutions, but even to write them down in a closed form (apart from some special cases). However, all results obtained so far (in some particularly simple cases that we can deal with) are encouraging, since they seem to suggest that the quantum corrections just provide damping corrections to the classical, accelerated evolution, and tend to favor the transition to the phase described by the standard cosmological model.

The relevance of the quantum corrections suggests that the obstacles we encounter when we attempt to describe the transition could be surmounted by abandoning the classical, geometric approach, where we follow the space-time evolution point-by-point, moment-by-moment. Since the equations describing this evolution are not fully known, in general, it could be convenient to adopt the probabilistic approach of quantum cosmology which does not require full knowledge of all the intermediate evolutionary stages, but only of the initial and final states.

It is worth noticing here that, even in the cosmological framework based upon Einstein's equations, there are open issues which it seems appropriate to address with quantum mechanical methods. We may recall for instance that, within the standard inflationary scenario, the primordial Universe approaches a state of exponential expansion and constant curvature. Such a state, described by the de Sitter geometry, cannot have lasted indefinitely in the past (see Chap. 1). Hence, we cannot avoid facing the problem of how this state might have emerged.

A possible solution to this problem was suggested independently during the 1980s by some Soviet cosmologists (Alexander Vilenkin, Andrei Linde, Valery Rubakov, Yakov Zeldovich, and Alexei Starobinski). Their solution relies upon the idea that the initial de Sitter state may emerge "from nothing", i.e., it may be spontaneously produced from the vacuum thanks to an effect called quantum tunneling. The tunneling effect is a well-known process in elementary particle physics, where a particle, represented by a quantum mechanical wave, is able to overcome a potential barrier even if its energy is inadequate at the classical level. (Very naively, it is as if a cyclist, who does not have enough energy to climb a small hill, unexpectedly finds a tunnel at the bottom of the hill that allows him to get through.)

In a cosmological setup, the description of the birth of the Universe in terms of the tunneling effect requires the introduction of a peculiar infinite-dimensional space, the so-called superspace, whose points represent all possible geometric configurations of the

Universe. For practical reasons it is possible to use a reduced space, dubbed mini-superspace, and characterized by a finite number of dimensions (associated, for instance, with the radii of a spatial section of the Universe measured along the different spatial axes). The motion of a wave from one point to another of this mini-superspace represents the transition of the Universe from one geometrical state to another, and is governed by the so-called Wheeler–DeWitt equation, named after two theoretical physicists (John Archibald Wheeler and Bryce DeWitt) who first proposed it in the 1960s.

The Wheeler–DeWitt equation is the exact analogue of the Schroedinger equation of ordinary quantum mechanics, the only difference being that its solutions, instead of describing the possible values of the position and momentum of a given physical system (for instance a particle), represent the possible geometrical states of the Universe. A Universe described by the Wheeler–DeWitt equation thus becomes a fully quantum mechanical Universe, subject to all possible quantum effects. We know, for instance, that the so-called second quantization of the Schroedinger wave function leads to the formalism of quantum field theory, where it is possible to describe the creation and annihilation of particles. Similarly, quantization of the Wheeler–DeWitt wave function gives rise to the so-called third-quantization formalism, where it is possible to describe the creation and annihilation of universes.

By an appropriate choice of initial conditions, it is possible in particular to find solutions to the Wheeler–DeWitt equations describing the birth of our Universe as a tunneling process, thus providing a solution to the classical problem of the origin of the inflationary de Sitter space. One finds that if the state of the Universe after the tunneling process is described by the de Sitter geometry, and is thus characterized by a constant Λ representing the vacuum energy density, then the bigger the value of Λ, the higher the tunnelling or transition probability. In this way, the Universe is created just in the appropriate inflationary state, which does indeed require a high enough value for the parameter Λ (also called the cosmological constant).

Barring a number of formal and technical problems, the most unsatisfactory feature of this scenario is probably the fact that the initial conditions for the tunneling process are to be chosen ad hoc, since they are not unique. There are also arguments supporting the choice of different initial conditions (as discussed, again during the 1980s, by other theoretical physicists including James Hartle and Stephen Hawking), which lead to different scenarios. The reason for this arbitrariness (dubbed the boundary condition problem) is rooted in the fact that within standard cosmology the final state – i.e., the cosmological configuration we aim to obtain – is well known, whereas the initial state is completely unknown. Indeed, the very name of the tunneling process, "tunneling *from nothing*", already automatically stresses the lack of knowledge about the initial state. The standard classical theory is not helpful at all, since it just predicts the Big Bang singularity as initial state, i.e., the state that the quantum mechanical approach would like to avoid.

Within the self-dual pre-Big-Bang scenario, the situation is radically different. The initial state, assumed to be the perturbative vacuum of string theory, is completely known, fully justified, and fully appropriate to be described – in the low energy regime – by the Wheeler–DeWitt wave function. Given the initial state, the computation of the transition probability towards the final state, i.e., the current Universe, is no longer arbitrary.

It is therefore interesting to note that, by computing the probability that a transition occurs between the perturbative string vacuum and a post-Big-Bang Universe equipped with a cosmological constant, the outcome is quite similar to the result obtained in standard cosmology assuming the validity of the "tunnelling from nothing" scenario (as shown by Gabriele Veneziano, Jnan Maharana, and the present author). This could suggest that the ad hoc prescription for the boundary conditions, needed to obtain the tunnelling effect, somehow simulates the presence of the perturbative vacuum as initial state. It would then be more appropriate to talk about "tunneling from the string perturbative vacuum", rather than "tunneling from nothing". Figure 8.1 provides a qualitative representation of this result.

There is, however, a conceptual difference between string cosmology and standard cosmology. The quantum mechanical transition from pre-Big-Bang to post-Big-Bang described by the Wheeler–DeWitt equation, in a two-dimensional mini-superspace where the coordinates are represented by the spatial radius of the Universe and the dilaton, does not correspond to a tunnelling effect, but rather to a quantum reflection effect.

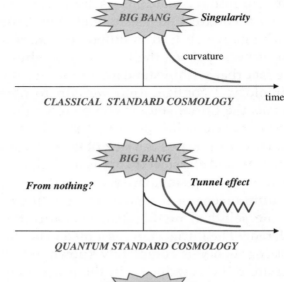

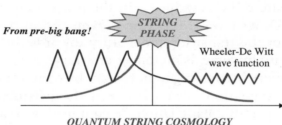

FIGURE 8.1 *From top to bottom*, a qualitative comparison between classical standard cosmology (characterized by the singularity), quantum standard cosmology (with appropriate boundary condition for the tunnelling effect), and quantum string cosmology (according to the pre-Big-Bang scenario). The *zigzag curves* represent the Wheeler–DeWitt wave function, which is asymptotically oscillating before and after the transition, while it is exponentially decaying in the region of the tunnel effect. Note that both the classical and quantum standard scenario are characterized by the Big Bang singularity, and cannot be extended beyond it. In string cosmology the singularity is replaced by a high-curvature string phase

The reflection of a particle from a barrier is part of our everyday experience and – unlike the tunneling effect – it is obviously also present in the context of classical mechanics. However, within quantum mechanics, there also exists the possibility of a new, truly "quantum", reflection effect relative to particles which, despite having the energy required to climb over the barrier, are instead pushed backwards, in a completely unexpected way from a classical perspective. It is just as if a bullet, shot from a gun against an easily perforated target – like paper, for instance – were to bounce back instead of passing through the target, an impossible effect in the context of classical mechanics, but not within quantum mechanics.

In the pre-Big-Bang scenario, the barrier is represented by the region of very high energy and curvature that divides the post-Big-Bang phase from the singularity in mini-superspace. Without the presence of quantum effects, the Wheeler–DeWitt wave would easily overcome this barrier, ending up in the infinite-curvature pit that leads to the singularity (as predicted by the classical cosmological dynamics of the low-energy solutions). Instead, thanks to the possibility of quantum mechanical reflection, there is a finite probability that the wave reaching the barrier will be bounced back, thus describing a Universe that moves towards a quiet end after having happily reached the standard, post-Big-Bang evolutionary phase (see Fig. 8.2).

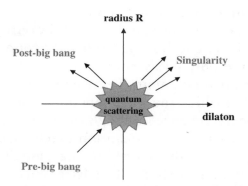

FIGURE 8.2 Qualitative sketch of the quantum transition from pre-Big-Bang to post-Big-Bang, represented as a quantum mechanical reflection of the wave function in a mini-superspace whose coordinates correspond to the dilaton and to the spatial radius of the Universe. The incident wave (*bottom left*) describes the initial evolution of the Universe from the string perturbative vacuum towards the high curvature regime. Part of this incoming wave is not stopped by the barrier and is classically transmitted to the region of ever increasing dilaton, running towards the singularity (*top right*). Another part is reflected to the region of decreasing dilaton and standard post-Big-Bang evolution (*top left*)

The transition from the pre-Big-Bang to the post-Big-Bang regime, which somehow represents the birth of the Universe in the form we are currently observing it, can then be described according to quantum string cosmology as a process of scattering and reflection of the Wheeler–DeWitt wave function in mini-superspace. It may be noted that, whereas in the standard cosmological scenario quantum effects are required for the Universe to *enter* the inflationary regime, in string cosmology such effects are required for the Universe to be able to *leave* the inflationary pre-Big-Bang regime, and enter the phase of standard evolution. Instead of a tunneling effect, there is a reflection process. The conceptual differences are evident, but the methods and the formalism are the same.

It must be pointed out, however, that if the transition occurs via either a tunneling or a quantum reflection process, then the final oscillation amplitude of the wave function turns out to be significantly reduced with respect to the initial amplitude (see Fig. 8.1). This means that the transition probability is quite small, i.e., the transition mechanism is not effective. Then, it would be hard for the Universe to exit from the pre-Big-Bang phase and to fulfill its standard evolution up to now.

However, the above mechanism is not the only way the transition can proceed (more technically stated, the tunneling/reflection process is not the only decay channel of the string perturbative vacuum). There are other, more effective processes (studied by the present author) in which the wave function, instead of being suppressed, is strongly enhanced through the high-curvature regime. For such processes there is a quantum mechanism that acts inversely with respect to the one producing tunneling (it is in fact called the anti-tunneling effect). This latter effect is quite similar to the one described in Chap. 6, which amplifies the quantum fluctuations, with a subsequent production of particle pairs from the vacuum.

The crucial difference with respect to the process of Chap. 6 is that the oscillations now being amplified are those of the Wheeler–DeWitt wave function, representing the evolution of the Universe. Hence, the process can once again be described (in the context of third quantization of the Wheeler–DeWitt wave function) as a process of pair creation. However, the resulting pairs are not particles, but rather *pairs of universes*, directly produced from the string perturbative vacuum, i.e., from the initial state of the pre-Big-Bang scenario (see Fig. 8.3). In each pair, one of the created universes is absorbed by the singularity (falling in the portion of mini-superspace where curvature and dilaton growth is unbounded, see Fig. 8.2), and disappears from our present experience. The other Universe evolves towards the opposite, low-curvature regime, thus entering into the post-Big-Bang phase, eventually to approach the current regime.

In the same way as the particles emerging from the vacuum are produced in pairs and characterized by opposite physical properties (i.e., opposite charges, momentum, angular momentum, etc.), to avoid violations of conservation laws, the universes are also produced in pairs, and are characterized by opposite kinematic properties. One of the two universes expands while the other shrinks. However, the shrinking Universe behaves as if it were traveling backwards in time with respect to the coordinate playing the role of time in mini-superspace.

It is well known, on the other hand, within the context of second quantization, that a particle moving backwards in time ought to be interpreted physically as an antiparticle, with opposite charge, moving forward in time. Thus, in a third quantization context, the shrinking Universe must be reinterpreted as an anti-universe which is expanding, and the anti-tunnelling process must be seen as pair

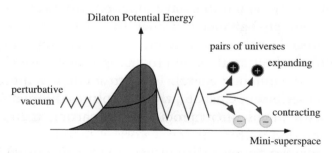

FIGURE 8.3 Schematic view of the quantum transition from pre-Big-Bang to post-Big-Bang represented as an anti-tunneling effect of the wave function, i.e., as a creation of pairs of universes from the string perturbative vacuum. The Wheeler–DeWitt wave function is amplified during this process. The opposite charges of the resulting universes are to be interpreted as opposite kinematic states, corresponding to either expansion or contraction in mini-superspace

production of universes and anti-universes, both expanding, one towards the singularity and the other towards the current low-energy regime. Unlike the quantum reflection process, such a process can be quite efficient, as long as the dilaton interaction can provide the potential energy required for the occurrence of pair production in mini-superspace (in Fig. 8.3 this potential energy is represented by the barrier that amplifies the wave function).

To conclude, we can say that in the framework of quantum cosmology the transition from the pre-Big-Bang to the post-Big-Bang phase can be described in probabilistic terms, even without any detailed knowledge of the kinematics and dynamics of the high-curvature, strong-coupling regime. However, the use of the Wheeler–DeWitt equation obtained from the low-energy classical description is still an approximate approach. The existence of a minimum length within string theory does indeed imply that, close to the region of maximum curvature, the equations for the classical fields are modified by corrections including the square, the cube, and all higher powers of the curvature (see Chap. 4). Those corrections could also modify the Wheeler–DeWitt equation and the geometry of the mini-superspace.

According to string theory, a fully exact (to all orders) and consistent description of the Universe in the quantum regime should probably abandon concepts such as fields and geometry, and rely solely upon the motion of strings and the feature of conformal symmetries. Actually (as pointed out in Chap. 4), it is precisely from these symmetries that modifications to the equation of general relativity arise. Hence, it is precisely from them that the possibility of avoiding the initial singularity of standard cosmology may originate (as suggested by various studies and many authors). It is therefore possible that present quantum cosmology models will be improved by future developments of string theory, and eventually by its completion within the framework of membrane theory and M-theory, a recently born theoretical framework whose development looks promising (see Chap. 10).

9. The Future of our Universe

One the most fascinating features of cosmological models – apart from their ability to describe the current state of the Universe and to trace its past history – is their ability to make predictions concerning future evolution. Obviously, the more accurate the knowledge of the current cosmological state, the more accurate and reliable such predictions become. As long as the number of observations increases, with progressively better experimental sensitivity, our knowledge of the current Universe is continually subject to revision and updates. Hence, predictions about the future may also need frequent revision and improvement.

In this chapter we focus on a discovery that can be counted amongst the most important made at the end of the last century, a discovery that has radically changed an already consolidated view, bringing a new perspective into our ordinary expectations about the future of the Universe: rather than slowing down as predicted by the standard model, the cosmological expansion is currently accelerating!

To explain why such discovery is so revolutionary and "explosive", and understand the subtle link that could exist between such acceleration and the pre-Big-Bang cosmology described in the previous chapters, it is wise to proceed step by step. Let us start by recalling that the standard cosmological model, using the equations of general relativity and the assumptions of homogeneity, isotropy, and perfect fluid sources (see Chap. 2), provides us with a very sharp description of the current Universe. It should be noted, incidentally, that the use of the classical theory of general relativity is certainly legitimate for describing geometrical configurations with a curvature much smaller than the Planck curvature (like the current Universe), since in that case the possible modifications of the theory due to quantum effects are completely negligible.

According to the equations of general relativity and the assumptions of the standard cosmological model, it is found, in particular, that the current evolution of the cosmological geometry – once the constant value of the spatial curvature has been fixed – is fully described by a unique time-dependent quantity: the spatial radius (or scale factor) $R(t)$. The time behavior of this quantity will tell us whether the future Universe will always be expanding (in a decelerated or accelerated fashion), or become static, or start to shrink, eventually collapsing towards a future singularity.

In order to solve the equations of general relativity, and therefore to be able to predict the future behavior of the cosmic space-time geometry, it is essential to have a precise knowledge of at least three parameters peculiar to the current cosmological state. For instance, (1) the current average energy density at cosmological scales, (2) the corresponding equation of state, i.e., the average pressure of the dominant cosmological sources, and (3) the current value of the average *spatial* curvature. Note that the third requirement does not refer to the full, space-time curvature, but rather to the curvature of the geometric manifold obtained by taking a spatial "slice" of the Universe at a given time.

It should be stressed that the three reference parameters could be different from the ones just mentioned. In the context of the standard cosmological model, in particular, the equation of state is fixed by assuming that the average current pressure is zero. Then, a measurement of the energy density alone is enough to determine the spatial curvature, which turns out to be positive, negative, or zero depending on whether the energy density is higher than, equal to, or smaller than a certain value called the critical density (about 10^{-29} g per cubic centimeter). However, in order to fix the critical density, and compare its value with the present density, we need to determine a third characteristic parameter of our epoch: the Hubble parameter H_0, the one controlling (to a first approximation) the present recession velocity of the galaxies.

Given H_0, the cosmological energy density, and the pressure (the latter assumed to be zero) as three independent parameters, the standard model can then unambiguously establish whether we live in a Universe that will remain forever in a state of decelerated expansion (critical density, zero spatial curvature), progressively approach the flat and empty Minkowski space (sub-critical density, negative spatial curvature), or stop expanding and subsequently recollapse (super-critical density, positive spatial curvature).

To test the standard model, and hence decide indirectly which fate among the various possibilities the Universe is going to realise, astronomers and astrophysicists have employed all their skills to obtain more and more accurate measurements of the observable quantities characterizing the current cosmological state. Those observational developments brought a first surprise about thirty years ago, one that forced us to consider a first modification of the simplest, original form of the standard model.

The standard model originally assumed that the type of matter that currently represents the dominant form of energy over cosmological scales should be made of atoms, and in particular protons and neutrons present in their nuclei, contained not only in the planets and stars but also (with great abundance) in the dust filling the galactic and intergalactic space. We may synthetically refer to this type of matter as baryonic matter, since the baryons are a class of "heavy" elementary particles, having the proton itself as the most fundamental and stable component.

Why should baryons represent the currently dominating component of the cosmic energy? The answer is simple. Since baryons are heavy particles, as the Universe becomes progressively colder their kinetic energy becomes negligible, i.e., they become non-relativistic, almost static particles which can on average be described in terms of a zero-pressure gas. Then, according to the Einstein equations, their energy density decreases with time as the reciprocal of the volume, i.e., the reciprocal of the third power of the spatial radius. On the other hand, the energy density of relativistic particles – hence that of radiation – decreases more rapidly, as the reciprocal of the fourth power of the radius (see Chap. 2). This implies that, as the Universe has expanded, the radiation energy density has been diluted much faster than the baryonic energy density, and today should have become a subdominant component of the fluid filling the Universe on cosmological scales.

This prediction is well confirmed by direct observations. Collecting together all the radiation and the relativistic particles currently observed, the result is that their energy density is about one ten thousandth of (i.e., 10^{-4} times smaller than) the critical density, the main contribution to this number coming from the cosmic electromagnetic background already mentioned in previous chapters. Summing instead all the baryons within the galaxies, the intergalactic dust, and all visible matter one gets an energy density which is about one hundredth of the critical value, hence one hundred times bigger than the radiation energy density. Therefore, it would seem safe to conclude that the Universe is presently dominated by a gas of non-relativistic baryonic particles with zero average pressure.

This quite simple conclusion, consistent with the standard model, has nevertheless been blatantly contradicted by other astrophysical observations. In fact, if the main gravitational source is a gas with zero pressure, then the Einstein cosmological equations tell us that the ratio between the density of this gas and the critical density is proportional (with a factor of two) to a kinematical quantity called the deceleration parameter, whose value depends only on the acceleration of the Universe. Now, the measurements of this parameter, despite large errors and uncertainties, have provided us with a value which, as early as the 1970s, was known to be of order one, thus implying that the matter density is of the same order as the critical density. But the baryons – as stressed above – have a density which is only one hundredth of the critical value, so they cannot be the current dominant form of energy!

This sort of enigma, also called the missing mass problem in the 1970s, can be solved by assuming that the dominant form of energy in the present Universe is made of non-baryonic, non-relativistic matter which is invisible to optical observations (or to other types of direct electromagnetic detection), and whose effects are of a purely gravitational nature, e.g., through its influence upon the space-time curvature and the expansion rate of the Universe. Such a cosmic fluid has been dubbed dark matter. The introduction of this matter component undoubtedly explains not only the discrepancy between the observed values of critical density and the baryon density, but also other important astrophysical observations. As an example, the presence of dark matter in the galactic halo explains why the stars rotate around the galactic center more rapidly than is predicted by the standard gravitational theory under the assumption of an empty interstellar medium.

Over the last twenty years, the dark matter hypothesis has inspired a very great deal of research, both theoretical and experimental.

From a theoretical perspective various attempts have been made to develop physically "acceptable" models of dark matter. Indeed, the crucial question is: If it is not baryonic, what kinds of particles are the basic components of dark matter? A number of models have been studied, with conventional non-baryonic particles (e.g., neutrinos), with more exotic particles (axions, dilatons, etc.), and also with supersymmetric particles (photinos, gravitinos, etc.). On the experimental side there have been attempts to detect this kind of "invisible" matter either directly or indirectly, exploiting various types of observation. For instance, astrophysical observations measuring the micro-lensing effect, i.e., the microscopic amplification of light rays emitted by stars due to the gravitational field of dark matter. Other dark matter searches are based on underground observations, i.e., observations carried out with particle detectors located underground, in order to remove other signals (like those produced by cosmic rays) that could mask the effects due to the interaction of the dark particles with the detector.

Today, the issues concerning the identification and the direct observation of dark matter have not yet been fully clarified. However, it is legitimate to say that, thanks to the introduction of dark matter, a version of the standard model was developed and improved over the last twenty years of the last century, a version that until recently seemed to be able to explain all the observations concerning the present state of the Universe, and even to match quite well with the predictions of the inflationary scenario. In fact, the presence of dark matter allowed the computation of the small anisotropies of the cosmic radiation in good agreement with the precision measurements that have been made, starting with the COBE experiment, since 1992.

But in this situation of idyllic agreement between theory and observation, a storm was already on the way. In the meantime, a set of data and observations were being collected,[1] revealing a rather sharp contrast with this scenario. These contradictions were something of a bomb shell during the years 1997–1998, producing a crisis with the standard dark matter scenario. In fact, according to those observations, not only baryons, but even dark matter (that

---

[1] S. Perlmutter et al.: Nature **391**, 51 (1998); A.G. Riess et al.: Astron. J. **116**, 1009 (1998).

should in any case be present within all galaxies with a density one hundred times greater than the baryonic density) cannot account for the cosmic component that is dominant today on large scales!

What are the revolutionary observations that lead us to modify the standard assumptions about the current Universe so drastically? They are similar to (but more accurate than) the ones that, almost a century ago, allowed us to discover the Hubble law, linking the distance with the redshift of the most distant objects we are able to observe. In particular, these observations concern supernovae, huge nuclear explosions that mark the endpoint to the "ordinary" life of highly massive stars, transforming them into neutron stars (and possibly black holes).

Why do we focus on supernovae to measure the geometry of the Universe? There are two main reasons. First of all because they are highly intense sources of light, and hence can be observed at very great distances; some of them are so far from us that the light emitted from their explosion reaches us after traveling for a time comparable with the present Hubble time $1/H_0$, whence their light reaches us from regions of space located very near to the present horizon. The other reason is that their intrinsic luminosity (also called absolute luminosity), i.e., the amount of light (and energy) they emit per unit time, is relatively well known, and should only depend on the considered type of supernova. (The particular class of supernovae analyzed for those observations is called type Ia.)

Actually, knowing their absolute luminosity, it is possible to express their apparent luminosity, i.e., the amount of light that reaches us, as a function of their distance or, better, as a function of the redshift z suffered by the light of those supernovae during its journey, due to the expansion of the Universe (see Chap. 2). The observed data can then be used to construct what is known as the Hubble diagram, which provides the apparent luminosity of the observed supernovae as a function of their redshifts. And herein lies the crucial result from these observations.

In fact, if we compute the redshift as a function of the distance using the equations of the standard model, we obtain a relation which is linear only to the first approximation – for z sufficiently less than one – in agreement with the well-known Hubble law. In general, the relation between redshift and distance is nonlinear, and depends on the geometry under consideration. The geometry itself depends in turn on the amount of matter and energy present, and on how these sources warp the space-time. Thus, if we draw a graph of the apparent luminosity (or rather the apparent magnitude, to use the astronomers' jargon) versus z, we have a number of different possible curves, each of them corresponding to different possible models of the Universe with different pressure and energy-density contents.

If we now compare these curves with the data relative to supernova observations, plotting the magnitude against the redshift, the result is that supernovae tend to align themselves along curves corresponding to a geometry generated by sources with negative pressure! Hence, the currently dominating energy density cannot be either baryonic matter or non-relativistic dark matter, since both of them give zero cosmological pressure. There must therefore be, at the cosmological level, a hitherto unknown form of energy with negative pressure, suggestively called dark energy.

The most recent observations, obtained by combining the latest supernova data[2] and the CMB data of the WMAP satellite, seem to confirm that the current value of this dark energy density represents roughly seventy per cent of the total cosmological energy density, while the remaining thirty per cent almost completely consists of dark matter, except for the tiny contribution due to baryons (of the order of one per cent) and the extremely small contribution associated with radiation (of the order of one ten thousandth). It is indeed fascinating to find that our Universe is almost completely made up of components that are invisible (apart from their gravitational effects).

The above observations also tell us how large the dark energy pressure can be. Taking the ratio between the pressure and the energy density, we find that the result must be a negative number, very close to the value -1. And this yields to another very interesting outcome, already mentioned at the beginning of this chapter: the current Universe, dominated by this dark energy, must be expanding with a positive acceleration.

Indeed, according to Einstein's equations, the acceleration with which the spatial radius of the Universe changes with time can be obtained by summing the contributions of the energy

---

[2] P. Astier et al.: Astron. Astrophys. **447**, 31 (2006).

density and the pressure, and flipping the sign of the final result. The sign flipping is due to the fact that ordinary gravity is an attractive force, and its action tends to slow down the expansion, producing a negative acceleration, i.e., a deceleration. However, if the pressure is negative, its contribution with the sign flipped corresponds to a repulsive force, producing a positive acceleration and a continuous increase in the expansion velocity.

The pressure contribution to Einstein's equations, on the other hand, enters with a multiplicative factor of three compared with the contribution of the energy density (because the pressure of an isotropic fluid is equally distributed along three spatial dimensions, while the energy density is associated with the time dimension, which is unique). If the ratio between pressure and energy density is very near to -1, i.e., if their intensities are roughly equal in modulus, as suggested by the data, it is then evident that the pressure "wins" against the energy density, producing an overall repulsive force which leads to an accelerated cosmic evolution.

The question which arises naturally, at this stage (and which provides the link with string cosmology) is the following: What type of matter or field does this elusive dark energy correspond to? Is it produced by some known particle, or is it a more exotic effect emerging only on very large scales, i.e., at the cosmological level? In other words, what is the greater part of our Universe (almost seventy per cent) made of?

Despite the fact that the scientific research in this field started only very recently, there are already possible (and even plausible) answers to these questions. The first, and historically most natural, candidate to play the role of dark energy is undoubtedly the so-called cosmological constant Λ, a term introduced by Einstein in his equations just to simulate a "cosmic repulsion". According to modern quantum field theory this term is usually interpreted as the vacuum energy density, i.e., as the energy due to the sum of all the microscopic oscillations that the quantum fields must have, even in their ground state (i.e., in their lowest energy level), due to the Heisenberg uncertainty principle. Thanks to these oscillations, even in the absence of any body or particle, the vacuum acquires an average energy density which is constant, has a negative pressure (equal in modulus to the energy density) and can, like all forms of energy, generate a cosmic gravitational field.

The presence of a cosmological constant thus provides an explanation for the observed cosmic acceleration which works quite well phenomenologically. Indeed, the first supernova data were interpreted as evidence for a non-zero cosmological constant with a currently dominating energy density. However, there are serious formal and conceptual problems associated with this simple explanation, and most researchers are now inclined to reject it.

In fact, the cosmological energy density that we observe turns out to be considerably smaller than the typical vacuum energy computed using the current models of elementary particle physics. Recall that the mass density of the dark energy has to be of the same order as the critical density, i.e., about 10^{-29} g per cubic centimeter. Why is it so small? It would be easier to explain, using a symmetry principle, if the cosmological constant were exactly zero. Instead, a small but non-zero value leads to a fine-tuning problem: an extremely accurate and unnatural adjustment of the parameters of the theory is required to obtain this value, and this issue has not yet been resolved in a fully satisfactory way.

Another open issue concerns the fact that the current dark energy density has a value quite similar to the energy density of dark matter. Actually, if the dark energy density is due to a cosmological constant, its value does indeed remain *constant* in time, always fixed at the value we now observe. The dark matter density, on the other hand, decreases in time, because it is inversely proportional to the expanding cosmological volume. Thus, its past value was bigger than the current value, while its future value will be much smaller. We are then led to the so-called problem of cosmological coincidence: Why are the dark matter and dark energy densities approximately equal only in the current epoch?

A plausible solution to those problems suggests that the dark energy is not represented, in the Einstein equations, by a cosmological constant, but rather by a time-dependent term associated with the energy of some cosmic fluid or field. The simplest field to be considered is then a neutral scalar field, without intrinsic angular momentum and charge, and self-interacting, i.e., equipped with an appropriate potential energy. Scalar field models that may be appropriate for representing dark energy effects were suggested in pioneering work by Bharat Ratra, James Peebles, and Christof Wetterich, well before the experimental discovery of the cosmic acceleration, and later studied by many astrophysicists (Michael Turner, Martin White, Robert Caldwell, Rahul Dave, Paul Steinhardt, Ivaylo Zlatev, Li-Min Wang, and others).

This scalar field, however, should have properties so peculiar as to make it difficult to identify it with one of the scalar particles already present in the standard model. As anticipated in Chap. 2, a new term has indeed been appositely coined for this field: quintessence, a name which stresses its exotic character (it recalls the elusive fifth element conjectured by the ancient philosophers and still sought by alchemists in the Middle Ages). So what are the strange properties this cosmic field should have?

First of all, in order to have negative pressure, its energy density must be dominated by potential energy (the kinetic energy corresponds in fact to a positive pressure). Moreover, even if it varies in time, its total energy must be negligible compared with that of the other cosmological components for most of the past history of the Universe, in order not to affect all the successful predictions of the standard cosmological model (the formation of nuclei, the gravitational aggregation of non-relativistic matter, the subsequent production of galactic structures, and so on). It is only recently that its contribution ought to become relevant.

In addition, in order to play a dominant cosmological role in the present Universe, the mass of the particle associated with this field must be extremely tiny. In fact, the range of the corresponding force (which is inversely proportional to the mass) must be at least of the order of the Hubble radius of the present horizon, namely about ten billion light years. This value corresponds to a very tiny mass, about 10^{-66} g. (Recall that the electron, one of the lightest known particles, has a mass of about 10^{-27} g.)

Such a long range in turn generates other problems. If this particle carries a force among macroscopic bodies over such great distances, why has it not been found in any lab experiments so far performed? An explanation could be that the force is extremely weak, much weaker than all forces so far discovered, hence even much weaker than the gravitational force present in ordinary matter. Or maybe most matter is "neutral" with regard to the type of force carried by the quintessence field, and thus unable to feel it.

On the other hand, in order to solve the cosmic coincidence problem, the quintessence field should interact with dark matter,

at the cosmological level, with a force whose intensity should be almost equal to the intensity of the gravitational field (as suggested
by Luca Amendola for scalar-type models of dark energy, and by
Luis Chimento, Alejandro Jakubi, and Diego Pavon for fluid-type
models). To avoid contradictions with existing gravitational experiments, it therefore seems necessary to endow quintessence with
a further peculiar property: the possibility of different couplings
to different kinds of matter. In particular, the couplings should be
stronger in the case of the (still unknown) dark matter particles,
while they should be much weaker in the case of the known particles (protons, neutrons, etc.) composing ordinary matter.

It is worth noting that all the properties just mentioned as typical of the quintessence field can be satisfied by a scalar field which
has not been introduced ad hoc to explain the current astronomical
observations, but which must necessarily exist in a fundamental
theory like string theory: the dilaton field. Even in the context of
the pre-Big-Bang scenario, where the initial dilaton is massless and
has negligible potential energy, there are in fact quantum effects
able to generate a dilaton potential as soon as the Universe enters the strong coupling regime. Such a potential is known to go
rapidly to zero at weak couplings, but its behavior in the opposite,
strong coupling regime is not yet well understood. The cosmic
coincidence problem can in this case be explained in two ways,
depending on the behavior of the potential at strong couplings.

In fact, as repeatedly stressed in the previous chapters, the crucial property of the dilaton field is that it determines the strength of
the various natural forces (including the gravitational field). During
the phase preceding the Big Bang, the dilaton is subject to a rapid
and intense variation that brings the strengths of all forces from
initial values that are almost zero to final values approximately
equal to what is currently observed. It follows that during most of
the standard cosmological phase following the Big Bang, up to the
present epoch, the strengths of the various forces must have been
kept stable, fixed at nearly constant values. (Any possible variation,
if it exists, is so small as to have escaped unambiguous detection.)
Hence, the dilaton potential generated in the post-Big-Bang phase
must have the appropriate form to guarantee such stabilization, to
avoid contradictions with present observations. This may happen
in essentially two ways.

A first possibility is that the potential develops a series of local minima when approaching the strong coupling regime and that the dilaton, during the phase of post-Big-Bang evolution, is "trapped" inside one of those minima (exactly like a ball reaching the bottom of a hole, see Fig. 9.1). After some oscillations backward and forward, the dilaton "falls asleep" and, remaining fixed at a value ϕ_0, it also fixes the strengths of all forces at the values that we now observe.

In this scenario (studied by the present author) nothing else happens as long as the Universe is radiation dominated. However, as soon as the Universe enters the phase dominated by matter (see Chap. 2), the dilaton gets a kick that somehow wakes it up, tending to push it away from the equilibrium configuration. The kick is due to the fact that the dilaton couples to both the energy density and the pressure of the cosmic fluid sources, and that this coupling acts as a force in the dilaton equation of motion. In the radiation case, energy density and pressure compensate to give a null total

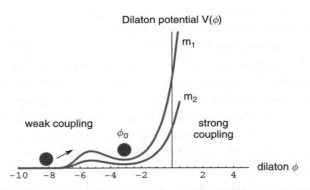

FIGURE 9.1 Possible plot of the dilaton potential for two different values of the dilaton mass, with $m_1 > m_2$. The potential goes to zero in the weak coupling regime, and has a local minimum at the beginning of the strong coupling regime, where the dilaton is trapped for the whole duration of the radiation era. Note that the amplitude of the potential depends on the mass and that, as the mass decreases (*lowest curve*), the depth of the potential well decreases too, so that it becomes progressively easier for the dilaton to escape from the equilibrium position ϕ_0. Hence, the dilaton mass has to be large enough to avoid being shifted away from the minimum, and small enough to correspond to a small potential energy which becomes significant only in late cosmological epochs

effect, while in the matter case the pressure is zero, compensation is impossible, and a force appears which tends to accelerate the dilaton.

If the dilaton were too light and too strongly coupled to matter, it could escape the potential well, and run back towards the large negative values typical of the pre-Big-Bang phase, dragging the coupling constants of the various forces towards progressively smaller values. As this did not happen, it follows that the dilaton is sufficiently heavy to remain confined at the bottom of the potential well (see Fig. 9.1). The dilaton mass, on the other hand, also controls the intensity of the dilaton potential and, in particular, its minimum potential energy, which should become the dominant source of the present acceleration. Hence, the mass must be sufficiently light to correspond to a potential which does not affect the Universe's evolution too early on. These two opposing effects leave us with only a limited range of values for the mass and the potential energy of the dilaton, thus alleviating – if not completely solving – the problem of explaining why the value of the dark energy density is such that it becomes dominant just at the current epoch.

An alternative solution to the coincidence problem (suggested by Federico Piazza, Gabriele Veneziano, and the present author, and subsequently studied by Luca Amendola and Carlo Ungarelli) can be obtained even if the dilaton does not remain trapped at a minimum of the potential, but keeps running toward higher and higher values, even during post-Big-Bang epochs. Of course, this is possible provided that the potential does not represent an insurmountable obstacle to the motion of the dilaton: after reaching a maximum, the potential must decrease towards zero, not only at large negative values of the dilaton (typical of the pre-Big-Bang regime), but also at large positive values of the dilaton, as in the example shown in Fig. 9.2.

Even in this case the strength of the various forces and couplings can be stabilized at constant finite values, with the difference (from the previous scenario) that the equilibrium position of the dilaton is now located at infinity (instead of being inside a potential well). In this case, as the dilaton grows towards higher and higher values (and, together with the dilaton, the bare value of the string coupling parameter also grows), the generated quantum

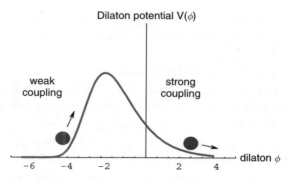

FIGURE 9.2 Possible plot of a non-perturbative dilaton potential which is smoothly decreasing in the strong coupling regime. In this case the dilaton, starting from the extremely large negative value of the initial state of the pre-Big-Bang scenario, can increase unconstrained towards arbitrarily large positive values

corrections become stronger and stronger, but they tend to saturate, i.e., to compensate one another, in such a way as to fix all couplings at a final constant value. And it is just within such a compensation mechanism – which becomes more and more efficient as the dilaton grows, i.e., as time goes on – that we can find a key for understanding the cosmic coincidence. As the dilaton couples "non-universally" to the various kinds of matter (see Chap. 7), it may happen that its coupling to ordinary matter becomes progressively smaller (and eventually negligible), while the coupling to the dark matter particles tends to stabilize at a higher value. Due to this effect, the Universe tends to evolve towards a final regime where the dark matter and the dilaton (which represents the dark energy) interact strongly together, while ordinary baryonic matter is decoupled.

Thanks to this coupling, the dilaton energy density, even if initially quite small, is progressively "dragged" towards the dark matter energy density, until a final regime is reached where the two densities are of the same order of magnitude, and evolve in time in the same way (see Fig. 9.3). This final "freezing" regime, reached at large enough values of the dilaton, is thus characterized not only by the stabilization of the various coupling strengths, but also by the stabilization of the dark energy over dark matter density ratio. In addition, if the dilaton potential is sufficiently strong, the cosmic

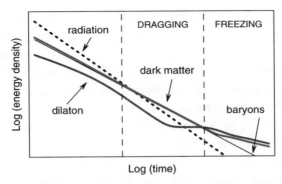

FIGURE 9.3 The figure shows on a logarithmic scale the time dependence of the radiation energy density (*dotted curve*), baryon energy density (*thin solid line*), dark matter and dilaton energy density (*thick solid lines*). They are obtained in a model where the dilaton, at late enough times, becomes strongly coupled to dark matter (but not to ordinary baryonic matter) and where, after a "dragging" phase, the Universe reaches a final accelerated "freezing" phase, in which the dark matter and dilaton energy densities are of the same order of magnitude and evolve in time in the same way. Note that, in this final regime, the baryon matter density is diluted faster than the dark energy and dilaton energy densities

expansion in this regime becomes accelerated, just as presently observed. The cosmic coincidence can thus be explained, in this context, by assuming that today we are already inside (or very near to) the freezing configuration, whence we may already observe that dark matter and dark energy have energy densities of the same order of magnitude.

There are various important differences between this second dilatonic dark energy scenario (which we shall call the coupled quintessence scenario) and the first one (more similar to conventional models of uncoupled quintessence).

First of all, in the coupled scenario the dark matter density will always remain of the same order as the dark energy density in the future, whereas in the decoupled scenario, the dark matter density will progressively become smaller and eventually negligible with respect to the dark energy density. What is diluted in time, in the coupled scenario, is the ratio between the baryonic and dark matter densities. This occurs because the dark matter density, thanks to the coupling to the dilaton, does not decrease proportionally to the reciprocal of the volume, as happens to the baryon density, but much more slowly (see Fig. 9.3). Incidentally, this effect could even explain why baryonic matter is today so much less abundant than dark matter, in spite of having the same equation of state. As a further important difference, we should mention the fact that the accelerated regime of the coupled scenario may start at much earlier epochs than in the decoupled scenario.

We do not yet know which scenario is the most realistic, and we do not even know whether one of them does effectively correspond to the effects we observe. Future observations will tell us. In fact, rather fortunately, the differences between the various scenarios mentioned above are observable (at least in principle). Notwithstanding, we may safely argue that the dilaton represents (at least at the moment) a plausible candidate to play the role of the quintessence field, i.e., the dark energy that is accelerating the expansion of our Universe.

To conclude this chapter we note that the standard cosmological model, although it is effectively able to provide (probably for the first time in human history) an accurate and scientifically consistent quantitative description of the Universe and the physical processes determining its current state, cannot be extrapolated either too far backward or too far forward in time. Current observations point to the need for modifications at small times (as discussed in previous chapters) as well as at large times (as outlined in this chapter).

Probably, from a purely conceptual perspective, it would be desirable for such modifications not to be detached from one another, but rather that they should originate from a unique theoretical framework or model. String cosmology, and in particular the pre-Big-Bang scenario, may provide a positive response to this requirement thanks to the presence of the dilaton, which is something of a general factotum in string theory; not satisfied with the modifications to general relativity and the primordial history of the cosmos, it also seems able to determine the future of the Universe, progressively becoming the most relevant form of energy.

If this were true, then the future Universe would be characterized by a strict relationship with the pre-Big-Bang Universe, and the current experimentally observed acceleration would already represent an indirect confirmation of the primordial scenario described in previous chapters.

Finally, if the cosmological dynamics is so tightly controlled by the dilaton, it is also possible to envisage a new and interesting scenario for the future of our Universe. Let us suppose, in fact, that the post-Big-Bang growth of the dilaton does not continue forever up to infinity. At some point the dilaton is stopped because its potential – which we do not yet know, unfortunately – forces it to bounce back towards negative values, i.e., towards the weak coupling regime. This would reproduce initial conditions appropriate to a pre-Big-Bang phase, and the curvature could start increasing again, thus initiating a new cycle of self-dual evolution. With a suitable form of the dilaton potential this sequence of events could repeat itself an arbitrary number of times, eventually implementing a cyclic scenario (similar to that obtained in the context of the ekpyrotic model, to be illustrated in the next chapter).

10. Recent Developments: Brane Cosmology Scenarios

About ten years ago, at the end of the last century, we witnessed the appearance of two new players on the scene of physical and astronomical science. We are just beginning to appreciate their dramatic impact upon basic cosmology, and we are probably far from a full understanding of all their implications. One of these two players brings observational novelties concerning the present accelerated evolution of our four-dimensional space-time on a macroscopic scale. The other brings novelties (probably more speculative) concerning a series of improvements in our theoretical understanding of the dimensionality of our world, culminating in the discovery of a possible mechanism for the confinement of gravity within four space-time dimensions, and opening new perspectives for the geometric description of higher-dimensional universes.

Both novelties are rich in cosmological applications and consequences. The first, discussed in the last chapter, has drastically changed our expectations about the future evolution of the cosmos. The second, which will be the subject of this chapter, has paved the way to new possible scenarios for the primordial Universe, in particular, for the description of the phase preceding the Big Bang and for models of the Big Bang itself.

As already pointed out in previous chapters, there are indeed many deep reasons – arising in the context of modern unified theories of fundamental interactions – for believing that the world in which we live is characterized by a number of spatial dimensions greater than three (at least nine, according to superstring theory). If we accept this view, however, we are left with the problem of explaining why only three spatial dimensions seem to be accessible to our ordinary experience, and to physical exploration through the instruments provided by current technology.

The obvious answer to this question, until a few years ago, was that all physical objects of our world (ourselves included) are characterized by a higher-dimensional extension along all possible available spatial dimensions (more than three, in general). However, only three of these dimensions have expanded over macroscopic (and larger) scales, giving rise to the currently observed Universe. The remaining dimensions (also called extra or internal dimensions) have instead evolved towards an extremely small and compact size, so as to become effectively invisible to all experiments so far performed.

We may think, for instance, of a long thin electric wire suspended between two high-tension steel pylons. The wire is actually a three-dimensional object but, if we look at it from far enough away, it will look like a one-dimensional object, simply because the size of its transverse sections are much smaller than its length.

The above approach to the dimensionality problem is quite reasonable, and still valid. Recently, however, a different approach has been put forward, which seems to offer an alternative explanation of why we are unable to perceive the extra dimensions. For a quick anticipation of what will be discussed in detail later, we only remark here that, according to this alternative explanation, all components of our physical world (ourselves included) have a pure three-dimensional spatial extension, in spite of being embedded in a higher-dimensional spatial manifold. The extra dimensions are not constrained to have a very small thickness as in the previous case. In fact, they may have a large, possibly infinite, extension. They are invisible simply because all forces and interactions through which we can explore our physical world are strictly "confined" to three spatial dimensions, being unable to propagate in the additional "orthogonal" directions.

As a naive illustration of such a situation we may think of a small bug like an ant, climbing up a curtain hanging on the wall of a room. The space inside the room is certainly three-dimensional, and all dimensions are extended over distances of comparable size (very large with respect to the size of the ant observer). However the ant – being unable to fly out from the curtain – is confined to move only vertically and horizontally along the curtain: hence, it will effectively experience only two spatial dimensions, in spite of being embedded in a fully three-dimensional space.

This particular view of a higher-dimensional Universe might seem, after all, a rather obvious and trivial explanation of the observed dimensionality of our effective space-time manifold. But from a conceptual point of view it represents a truly innovative result, the fruit of the most recent progress in theoretical physics, and full of important consequences. Hence, it seems appropriate to present here a detailed explanation of how (and why) we are led to such a higher-dimensional scenario.

To this end, let us take a step backwards, returning to the starting point of this chapter, i.e., to the fact that we expect to live in a Universe with more than four space-time dimensions. What are the physical motivations leading us to such (in principle unnatural) expectations?

The motivations are at present only of a theoretical nature. In fact, it is fair to say that we do not yet know of any direct experimental result forcing us to consider the existence of extra (either space-like or time-like) dimensions. The theoretical motivations, on the other hand, are certainly not a recent issue. They have a long history, starting almost a hundred years ago with the work of two theoretical physicists, Theodore Kaluza and Oskar Klein.[1] At that time, stimulated by the success of general relativity (which provided an elegant geometric description of all gravitational forces), many theoretical physicists were trying to incorporate not only gravity but also the other known fundamental interaction, i.e., the electromagnetic interaction, into the space-time geometry.

The revolutionary proposal of Kaluza and Klein was to generalize the four-dimensional space-time of Einstein's theory by adding a fifth dimension, of space-like character, and to interpret the additional degrees of freedom of that extended geometry as quantities directly related to the electromagnetic interactions. In this context, for instance, the extra components of the curvature were related to the electromagnetic field strengths, the moment along the fifth dimension was related to the electric charge, and so on. According to their model, now universally known as the Kaluza–Klein model, the new spatial dimension was wrapped onto itself (i.e., following the usual language, it was compactified) to form a microscopic circle, whose radius turned out to be fixed by Newton's gravitational constant and the fundamental unit of electric charge. The presence

---

[1] T. Kaluza: Sitzungsber. Preuss. Akad. Wiss. Berlin **1921**, 966 (1921); O. Klein: Z. Phys. **37**, 895 (1926).

of the electromagnetic interaction was then "explained", in that context, as a consequence of the symmetry associated with the coordinate transformations in the fifth dimension.

The five-dimensional model of Kaluza and Klein represented a consistent and successful unification of electromagnetism and gravity, free from problems affecting many other attempts at geometric unification (some of them proposed by Einstein himself). However, it was soon abandoned with the appearance in physics of new (strong and weak) interactions, active at the nuclear level, which seemed to be inconsistent with such a unification scheme.

The model was reconsidered, many years later, after the development of the so-called gauge theories, which associate with each interaction a well-defined symmetry group. By increasing the number of extra dimensions, and assuming an appropriate geometric structure for such dimensions (in general more complicated than a simple product of Kaluza–Klein circles), it is possible to obtain manifolds admitting a group of non-Abelian symmetries and reproducing the symmetry group typical of strong, weak, and electromagnetic interactions. In this way all fundamental forces of nature can be unified into a single (higher-dimensional) geometric description. The tiny volume of the extra, compactified dimensions, as in the original model of Kaluza and Klein, may prevent a direct experimental detection of those parts of space extending along the extra "internal" dimensions.

It should be noted, at this point, that such an effective unification scheme might be regarded as somewhat contrived, since the extra spatial dimensions have been introduced ad hoc (they are not a compelling prediction of the underlying theory). In addition, gravity is included in this unified scheme only as a classical interaction (differently from the other interactions, which can be consistently quantized). It is known, on the other hand, that a consistent quantization of gravity can be successfully implemented in a string theory context. It is thus quite remarkable that, within string theory, the higher-dimensional unification scheme of gauge interactions also finds its theoretical consecration, together with a potential phenomenological efficacy.

In fact, for a consistent description of the string motion at the quantum level, the string has to be embedded in an external space-time (the so-called target manifold), which has dimension D

(known as the critical dimension) greater than four. The quantum states associated with the discrete oscillation levels of the quantized string are thus represented by higher-dimensional fields, describing the forces present in such a higher-dimensional space-time. To obtain our physical four-dimensional world, on the other hand, the extra $D-4$ dimensions present in the model must be compactified. If the geometry of the extra compactified dimensions admits the appropriate symmetry group – more precisely, if its metric is invariant under the appropriate group of coordinate transformations, called isometries – then our reduced four-dimensional world automatically acquires the various gauge symmetries reproducing the forces that we currently observe.

We are thus led to the following important question: How many spatial dimensions are required by quantum string theory? The answer depends on the type of string we are considering.

The first string models, proposed during the 1970s, were based on the analogy with a classical vibrating object, and were formulated in terms of "bosonic" variables representing the coordinates of the string in the higher-dimensional external space-time. This type of string, called a bosonic string, can be consistently quantized in an external Minkowski space-time with a critical dimension of $D = 26$ (one time-like and 25 space-like dimensions).

This number of dimensions guarantees, for both closed and open bosonic strings, that the quantized theory is free from the so-called ghost problem, that is, the appearance of states of negative norm (i.e., of imaginary length in the space of states), also associated with negative probabilities (which are a mathematical nonsense). However, the bosonic string theory, quantized in $D = 26$ dimensions, cannot avoid the presence in its spectrum of states with negative squared masses (i.e., imaginary masses), called tachyons. These states describe particles that should always move faster than light, thus apparently violating the basic causality principles at the foundation of the modern quantum theory of fields and particles. It is probably appropriate to recall that no tachyon particle has ever been observed.

Hence, tachyon states must be eliminated from the quantum string spectrum. The most popular (and presently also more effective) way of doing this is to generalize the bosonic string model by assuming that strings vibrate not in ordinary space but in the so-called superspace, a virtual manifold spanned by a set of coordinates containing an equal number of bosonic and fermionic degrees of freedom. In this way the string model automatically becomes supersymmetric (see Chap. 3 for a definition of supersymmetry). More precisely, each bosonic coordinate determining the string position in the external D-dimensional space acquires a fermionic partner, transforming as a real spinor (i.e., as a massless, spin-half fermion) under coordinate transformations in the two-dimensional world-sheet surface spanned by the string (see Fig. 4.1), and transforming as a vector under Lorentz transformations in the external D-dimensional space.

In this way, one arrives at the so-called superstring theory, which eliminates not only ghost states but also tachyons from the physical spectrum. In fact, in a supersymmetric theory, the lowest allowed level in the squared mass spectrum has to be zero (negative eigenvalues are forbidden). In addition, a superstring model automatically introduces into the theory the fermionic variables required to represent the fundamental matter fields (quarks and leptons, whose various combinations can reproduce all forms of matter currently observed).

The consistency of superstrings with the basic principles of quantum mechanics and relativity (i.e., the absence of ghost and tachyon states in the spectrum) requires an external space-time manifold with $D = 10$ dimensions (one time-like and nine space-like dimensions). Three of these spatial dimensions correspond to our ordinary macroscopic space. The other six dimensions, wrapped onto themselves and confined to such a compact volume that they have so far escaped direct detection, are in principle enough to contain all the symmetries required to reproduce the interactions observed in our low-energy four-dimensional world.

We can say, therefore, that superstring models seem to provide not only a consistent framework for the quantization of gravity, but also a coherent and compelling scheme for a unified description of all fundamental forces of nature. Should this be confirmed, it would mean that two among the boldest and longest-sought goals of theoretical physics have been achieved in one shot.

There is a problem, however, due to the fact that the geometry of the compact extra dimensions is highly non-trivial. It may reproduce not only the particles and the interactions typical of our low-energy Universe, but also many other types of interactions which apparently have nothing to do with our world. This is the so-called landscape problem, arising from the many possible effective theories existing as a low-energy limit of the exact, high-energy string model, and due to our present ignorance of the mechanism driving our world to adopt a particular low-energy realization of string theory. This tends to weaken the unifying power of the theory, as the theory gives us too many models of low-energy interactions, and we have to select ad hoc the most appropriate one for a realistic description of nature.

In addition, we should recall here that superstring theory is not unique. There are indeed *five* possible superstring models, physically different from each other but nevertheless satisfying all required properties of quantum consistency. So which is the "right" model to describe our Universe?

This difficulty, unlike the previous one concerning the landscape, seems to have been solved. To understand how, let us first briefly introduce the five different types of superstring, starting with the so-called type II model, describing closed strings. In a closed superstring, the oscillations of its bosonic and fermionic components can propagate along the string either in the clockwise or in the counter-clockwise direction. Hence we have two possibilities, automatically defining two associated subtypes of the model, type IIA and type IIB. For their explicit definition we must introduce the notion of chirality, which we may think of as a vector pointing in a preferred direction in an appropriate virtual space.

In the type IIA superstring model, the oscillations of the fermionic fields propagating around the string in opposite directions are characterized by opposite values of chirality, i.e., they point in opposite preferred directions in the chirality space. In a type IIB superstring model, on the other hand, the fermionic oscillations propagating in opposite directions are characterized by the same chirality value, so that they cannot be distinguished by looking at the properties of the chirality space. These properties are not of purely formal character, as they have an important physical counterpart in the different particle and field contents of the two models.

Another superstring model is the so-called type I model, describing closed and open superstrings. The two types of string in this model are non-oriented, i.e., they are invariant under exchange of the ends of the strings. In other words, there is no preferred direction along the spatial path joining two ends of the string. Open strings, on the other hand, can carry charges (of all types) on their ends. The type I model (in contrast to the type II model) can thus contain in its spectrum the fields describing strong, weak, and electromagnetic interactions generated by the charges located on the ends of open strings. The fields associated with gravitational interactions (the graviton, the dilaton, etc.) are instead contained in the closed string spectrum.

Finally, there is the so-called heterotic superstring model, describing closed oriented strings in which only half the physical degrees of freedom are supersymmetrized (for instance, those associated with modes moving clockwise), while the other half keeps its bosonic properties, and it is quantized without fermionic partners. The procedure is consistent because, for closed strings, modes moving clockwise and counter-clockwise are decoupled, and can be treated independently.

The bosonic part of the heterotic string, on the other hand, can be consistently quantized in an external space-time with 26 dimensions, while the quantization of the supersymmetric part of the string requires only 10 space-time dimensions, as remarked previously. The additional $26 - 10 = 16$ spatial dimensions present in this model can then be compactified, to obtain an effective ten-dimensional theory. However, it turns out that the compactified dimensions are in principle compatible with two different symmetry groups, associated with different sets of higher-dimensional gauge fields and interactions. Hence, we have two possible models of heterotic superstrings, referred to as type HO and type HE. These two types, together with the other three types (I, IIA, IIB) give a total of five different models.

The five superstring models, apparently so different, are closely linked to each other through the action of the duality symmetries introduced in Chap. 3. In fact, using the appropriate combination of duality transformations, it seems possible to switch from one superstring type to another. In particular, changing the sign of the dilaton ϕ, i.e., inverting the string coupling parameter exp ϕ, and then switching from the strong coupling to the weak coupling regime, we may pass from a superstring model in which the perturbative approximation is no longer valid to the dual model in which this approximation is in fact valid.

The existence of such a duality network has suggested the conjecture that the five superstring models may simply represent five different approximate versions (valid in different regimes) of a unique, more fundamental theory, called M-theory. The complete formulation of such a theory seems to require one additional space-like dimension than superstring theory, and hence a space-time with $D = 11$ dimensions. What is presently known about M-theory is that, at low enough energy (and small enough curvature), it can be approximated as a supergravity theory – i.e., as a supersymmetric theory of gravity – in its maximally extended version, discovered by Eugene Cremmer, Bernard Julia, and Joel Scherk in the 1980s. (This does indeed require eleven space-time dimensions for its formulation.) At high energies, on the other hand, we have at present no precise information about M-theory so that, following a popular joke, we can say that the letter M of the name stands for mystery (or mother of all theories, or monster theory, along with other suggestions). But the more appropriate interpretation of the name is probably membrane theory, since M-theory describes, besides strings, the dynamics of extended objects like membranes.

To understand why membranes may naturally appear in the M-theory context we should recall here the important theoretical results obtained a decade ago by Petr Horawa and Edward Witten at the University of Princeton. They showed that the growth of the coupling parameter of a superstring (i.e., the growth of the strength of all interactions) can be equivalently described by adding a new spatial dimension to the space-time, and then gradually increasing the size of this dimension, following the growth of the coupling. In this way, we pass from ten to eleven dimensions, so that a string acquires a transverse extension. It becomes a two-dimensional object, i.e., a membrane, with a transverse size directly controlled by the strength of the coupling (see Fig. 10.1).

This means that the fundamental (one-dimensional) building-blocks of string theory can be interpreted, in the M-theory context, as two-dimensional membranes (or two-branes) embedded in an eleven-dimensional external space-time. In the limit in which the coupling strength of the theory goes to zero, the size of the eleventh

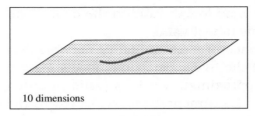

10 dimensions

vanishing coupling

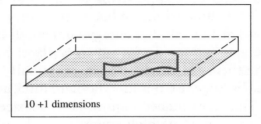

10 +1 dimensions

weak coupling

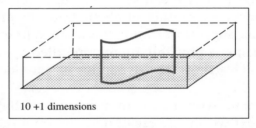

10 +1 dimensions

strong coupling

FIGURE 10.1 As the coupling strength increases, space-time acquires an additional space-like dimension, with a size proportional to the strength of the coupling. A string embedded in a ten-dimensional space-time becomes a two-dimensional membrane embedded in eleven dimensions

dimension becomes smaller and smaller, until the two-branes eventually degenerate to strings (see Fig. 10.1). It should be stressed that the eleventh dimension does not represent an additional direction along which a string can oscillate, so that there is no inconsistency between M-theory and the fact that a superstring requires precisely ten space-time dimensions.

The illustration of Fig. 10.1 refers to the case in which the coupling is growing, but the energy of the system remains low (and the space-time curvature remains small) with respect to the string scale. At higher energies, however, we may expect the theory to contain not only oscillating strings and membranes, but also higher-dimensional objects: three-dimensional extended bodies (called three-branes), four-dimensional extended bodies (called four-branes), and so on, up to nine-branes. In general, we will call an elementary object extended along p spatial dimensions a p-brane (a string is a one-brane, a particle is a zero-brane, and so on).

The important physical property of these higher-dimensional objects is that, in the weak coupling regime, they become very heavy, as their mass is proportional to the reciprocal of the coupling strength (and their mass grows with the number of spatial dimensions). However, it is obvious that, the heavier an object is, the more difficult it will be to produce it in a physical process, and the less important will be its contribution to a unified theory of fundamental interactions. Hence, strings are the most fundamental extended objects in the weak coupling regime.

In contrast, in the strong coupling regime, higher-dimensional branes become light and can be produced copiously, driving the Universe to a phase of brane-domination, typical of M-theory. Such a phase, triggered by the growth of the coupling predicted in the context of pre-Big-Bang models, could be the outcome of an epoch of pre-Big-Bang inflation, and could characterize the transition to the post-Big-Bang regime. But let us proceed step by step.

The first point to be stressed is that M-theory, as a theory of branes, suggests a new and revolutionary interpretation of the extra dimensions present in our Universe and required for the unification of the fundamental interactions. In fact, according to M-theory, the charges associated with the electromagnetic, strong, and weak interactions, as well as the fields of forces generated by these charges, may turn out to be strictly confined on a brane (as occurs, for instance, in the model studied by Horawa and Witten). This has suggested what is known as the braneworld scenario, based on the assumption that our three-dimensional world could be just a three-brane, embedded in an external eleven-dimensional space-time (also called bulk space-time), and that at least one of the extra space-like dimensions orthogonal to the brane – instead of being compactified and compressed to a very small (Planckian) volume – could have a very large extension (even infinite, in principle). Indeed, if all the interactions can only propagate on the brane, it becomes physically impossible to "feel" the extra orthogonal dimensions, no matter how "large" they are.

There is, however, an important exception to the confinement of the interactions on the brane. What applies to the gauge interactions – generated by the charges situated on the ends of open strings – does not necessarily apply to the gravitational interaction, a universal interaction associated with closed strings, and free to propagate in all spatial directions, even those orthogonal to the brane (see Fig. 10.2). Thus, using the gravitational force, we could in principle detect the presence of the extra dimensions (if they are

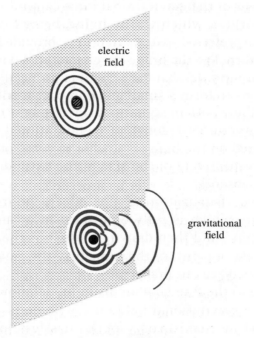

FIGURE 10.2 The grey two-dimensional surface represents a section of our four-dimensional braneworld, embedded in a higher-dimensional bulk manifold. The *small shaded upper disk* represents an electric charge located on the brane, and the surrounding *circles* represent the associated electromagnetic field. This type of gauge interaction is constrained to propagate only on the brane, and is not affected by the presence of the extra orthogonal dimensions. The *small black lower disk* represents a mass, and the surrounding *curves* the associated gravitational field. This field propagates not only on the brane but also outside it. As a consequence, the intensity of the gravitational force on the brane is weakened with respect to a pure four-dimensional world

large enough), in spite of our intrinsic nature as three-dimensional creatures, physically unable to get out of the brane. In fact, the gravitational force would spread along all available spatial directions, and its intensity on the brane would become weaker than predicted by the classical laws of Newton and Einstein. So how can one reconcile the possible existence of large extra dimensions orthogonal to the brane with the absence of any observed deviation from the laws of four-dimensional theories of gravity?

A possible answer to the above question has been provided recently by interesting work by two theoretical physicists, Lisa Randall and Raman Sundrum. They have shown that the long-range component of the gravitational force can indeed be confined on the braneworld in which we are living, being free to propagate only along three preferred spatial directions, provided that the extra dimensions external to the brane are characterized by an appropriate *curved* geometry. In that case there is no need to require the extra dimensions to form a small and compact manifold, as in the conventional Kaluza–Klein scenario. In fact, thanks to the action of the external curvature, the long-range gravitational field produced by a mass located on the brane is unable to come out of the brane itself, quite irrespectively of the size of the extra dimension (it is as if they were absent).

For a naive visualization of this effect, we may think of a thin metal plate which is fully embedded in a can filled with a very viscous paint, and then drawn out. Just as the plate remains covered by a coat of paint, in the same way the gravitational field tends to be "glued" to our brane.

According to the Randall–Sundrum model, however, the confinement of the gravitational field on the brane is a one hundred percent effective mechanism only for the massless, long-range component of the gravitational interaction. Just as in the case of the painted plate some heavier drops of paint can break off and drip away from the plate, in the same way the "heavier" components of the gravitational interaction (associated with a new type of massive graviton) could "drop away" from the brane, spreading out in the surrounding dimensions. If we could observe, and measure, such a small "leakage" of gravity, we could directly probe the existence of the extra spatial dimensions.

However, as stressed by the above analogy where the falling drops are the heavier ones, only the heavier components of the gravitational field, composed of massive particles, can propagate outside the brane. The exchange of these particles certainly induces corrections to the usual form of the gravitational forces, but such corrections have a short range. Summing up the contributions of all massive particles it turns out, in particular, that the corrections to the usual Newtonian potential between two gravitating bodies are proportional to the square of the curvature radius of the extra-dimensional space. If such a radius is sufficiently small (i.e., if the curvature is sufficiently large), then those gravitational corrections cannot be detected by present experiments, and the extra dimensions may remain invisible.

Concerning this point, we should recall that, according to recent experimental results, there is no observed deviation from the predictions of standard Newtonian gravity down to distances of about 0.1 millimeters.[2] This imposes an upper limit on the curvature radius of the extra dimensions, and thus an upper limit on the parameter that controls the strength of the gravitational interaction in the space external to the brane. Within the Randall–Sundrum model, the gravitational coupling constant in the external bulk manifold is determined by the product of the Newtonian constant G times the curvature radius of the bulk geometry. Hence, the bulk coupling constant may be larger than the usual, four-dimensional Newtonian constant, but not "too much" larger, if the size of the curvature radius is bounded.

The possibility of a bulk space with a large gravitational coupling parameter is indeed one of the main physical motivations at the heart of all higher-dimensional models with "large" extra dimensions; and not only if such dimensions are infinitely extended, as in the braneworld scenario of Randall and Sundrum, but also if they are compact, as in previous models proposed by Ignatios Antoniadis, Nima Arkani-Hamed, Savas Dimopoulos, and Gia Dvali. In fact, in our four-dimensional physical world, the strength of the gravitational force appears to be much weaker than the strength of all other forces active at the nuclear and subnuclear

---

[2] See, for instance, E.G. Adelberg, B.R. Heckel, and A.E. Nelson: Ann. Rev. Nucl. Part. Sci. **53**, 77 (2003).

level. This is at the origin of the so-called hierarchy problem: Why is there such a difference of strength among the fundamental forces of nature?

In models with large extra dimensions, the coupling strength of the bulk gravitational interaction can be much larger than in the case of four-dimensional gravity, and in principle similar to that of the other interactions, thus resolving (or alleviating) the hierarchy problem. The weakness of the gravitational forces that we experience in four dimensions would in this case be explained as a consequence of the higher-dimensional structure of our Universe, and the particular geometry characterizing the extra dimensions.

It is fair to say that such higher-dimensional scenarios are based on a number of assumptions, and are constrained by various phenomenological consequences. For instance, in order to generate the appropriate curvature of the extra-dimensional geometry, so as to confine four-dimensional gravity and possibly explain its hierarchical weakness, the Randall–Sundrum model requires the presence of a negative cosmological constant in the bulk space external to the brane. In other words, one needs a *negative* energy density for the higher-dimensional vacuum, a rather unconventional property which leads to a bulk geometry described by the anti-de Sitter metric (associated with a constant but negative space-time curvature). Among the new phenomenological consequences, we should mention, for instance, the fact that the brane (which is assumed to be static and rigidly fixed at a given position to a first approximation) could oscillate in the higher-dimensional space, thus generating massive scalar waves, possible sources of additional short-range corrections to the effective gravitational interaction on the brane.

If we make all necessary assumptions, and accept their possible phenomenological consequences, we may nevertheless formulate consistent models describing our world as a three-brane embedded in a higher-dimensional space-time. In this framework, we may develop a new approach to cosmology and a new perspective for the evolution of the primordial epochs. In particular, we may ask what happens if there is more than one brane, and if they can interact and collide with one another.

A possible answer to the last question has been provided recently by the collaboration of a group of astrophysicists and theoretical physicists, Justine Khoury and Paul Steinhardt at Princeton,

Burt Ovrut at Philadelphia, and Neil Turok at Cambridge. They have suggested that the collision of two (or more) branes might indeed simulate the Big Bang marking the beginning of standard cosmological evolution. The resulting cosmological scenario was termed "ekpyrotic", a name suggesting how our present Universe may emerge "from the fire" (i.e., from the outburst of radiation) produced by the collision of two branes. This scenario tries to explain the presence of the cosmic radiation background and of its temperature anisotropies through the process of brane collision, without resorting to a phase of standard inflationary expansion.

According to the ekpyrotic scenario, the Universe must therefore contain at least two three-branes, one (the "visible" brane) corresponding to our physical world, and another (the "invisible" brane) parallel to the first one, located at a certain distance along an extra dimension orthogonal to the branes. The scenario is inspired by the eleven-dimensional M-theory model of Horawa and Witten, and the two four-dimensional world-volumes spanned by the three-branes are domain walls representing the boundaries of a five-dimensional bulk-manifold (the remaining six spatial dimensions are assumed to be compactified on a much smaller scale). In the initial configuration the two branes are flat, parallel, and static, with no matter or radiation present in either brane.

There are two possible versions of the ekpyrotic scenario. According to the first version the bulk manifold contains a third floating brane, which is present from the beginning or which is formed spontaneously (at some subsequent time) thanks to the effects of quantum fluctuations. This bulk brane is attracted from the visible brane, and starts to move slowly towards it, moving faster and faster as the two branes get closer. The bulk brane is not perfectly flat like the other branes, but its geometry has small ripples due to the microscopic fluctuations of quantum fields in vacuum. When the two branes eventually collide, the kinetic energy of the bulk brane is fully transformed into matter and radiation, producing an outgoing flux of particles at very high temperature, giving rise to the the Big Bang. The visible brane, excited by the collision, heats up, gets curved, and eventually expands, reproducing our current Universe (see Fig. 10.3), while the small oscillations of the bulk brane would be the origin of the small anisotropies that we are currently observing in the cosmic background radiation.

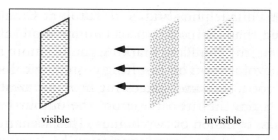

PRE-BIG BANG

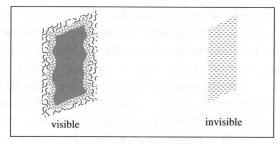

BIG BANG

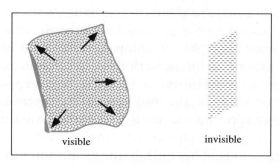

POST-BIG BANG

FIGURE 10.3 The two boundary branes are initially flat, cold, and empty. After the collision (simulating the Big Bang) our visible brane gets filled with radiation, gets hot, and starts expanding, giving rise to the phase of standard cosmological evolution

The second version of the ekpyrotic scenario is conceptually very similar to the first, with the difference that the colliding branes are now the two boundary branes. In that case the distances between the colliding branes coincides with the size of the extra "orthogonal" dimension, which is actually the eleventh dimension of the M-theory model, controlling the strength of all couplings.

The size of this dimension shrinks to zero during the pre-Big-Bang phase preceding the collision, and then bounces back to increase again when the two boundary branes separate before the collision. With an appropriate form of the effective potential controlling the dynamics of the interbrane distance, the branes could keep separating and colliding an (almost) infinite number of times, thus implementing the so-called cyclic scenario (proposed by Neil Turok and Paul Steinhardt soon after the formulation of the ekpyrotic scenario).

It is important to stress that there is a significant difference between the pre-Big-Bang phase of the ekpyrotic (or cyclic) scenario and that of the self-dual models described in Chap. 3. In spite of the fact that, in both cases, the initial configuration is flat, cold, and empty, the dilaton and the associated string coupling are indeed *decreasing* (with the interbrane distance) during the ekpyrotic phase preceding the brane collision – instead of growing, as in pre-Big-Bang models suggested by the duality symmetry. As a consequence, there is a possible technical simplification in the ekpyrotic scenario, due to the fact that the collision and bouncing of the branes occurs in the perturbative regime, where the string coupling becomes negligible. The initial state of the ekpyrotic scenario, on the other hand, has settled down in the strong coupling regime. Hence, such an initial state has to be "prepared" in some way. For instance (as already anticipated) by a previous epoch of growing dilaton, like the one typical of the self-dual pre-Big-Bang scenario.

When the coupling becomes strong, and the Universe enters the M-theory regime, there are other types of higher-dimensional objects which come into play, besides the branes marking the boundaries of the space-time manifold. In particular, there are the so-called Dirichlet branes (p-dimensional extended objects dubbed for short D_p-branes). Our four-dimensional world could just correspond to the hypersurface spanned by the evolution of a D_3-brane. These branes can interact among themselves, and their interaction can produce a phase of inflationary evolution (of conventional type), thus suggesting a primordial cosmological picture quite different from that of the ekpyrotic scenario.

Let us start by explaining what a D_p-brane is. To this end, we must recall that there are two types of string (open and closed), and that, when studying the propagation of an open string in a higher-dimensional space-time manifold, we have to specify what happens to the ends of the string, imposing appropriate boundary conditions. There are two types of condition:

- Neumann boundary conditions, if the ends of the string move in such a way that there is no momentum flowing through the boundaries.
- Dirichlet boundary conditions, if the ends of the string are held fixed.

If an open string is propagating through a background manifold with D space-time dimensions, then the position of the ends of the string can be determined in such a way as to satisfy Neumann conditions along $p + 1$ space-time dimensions, and Dirichlet conditions along the remaining $D - p - 1$ (spacelike) orthogonal directions. In this way the ends of an open string are localized on two p-dimensional hyperplanes at fixed positions (the two hyperplanes can also be coincident). Such p-dimensional hyperplanes are called D_p-branes.

It is important to stress that the ends of the open string are fixed along the Dirichlet directions, but can move freely along the (orthogonal) $p + 1$ Neumann directions, spanning the world-hypervolume of the brane. On the other hand, the string ends can carry charges, sources of Abelian or non-Abelian gauge fields. This gives us a natural implementation of the previously mentioned braneworld scenario, in which the fundamental gauge interactions are strictly localized on a $(p + 1)$-dimensional hypersurface, which is only a "slice" of the higher-dimensional bulk manifold in which the D_p-brane is embedded. In particular, D_3-brane could provide a model for our four-dimensional space-time.

In the context of superstring models, on the other hand, an extended object like a D_p-brane acts as source of an interaction which has a strength of gravitational intensity, and which is mediated by a totally antisymmetric tensor field of rank $p + 1$. In particular, D_3-branes can interact with one another not only gravitationally, but also through the exchange of a rank-four antisymmetric tensor field. Such an interaction, unlike gravity, is repulsive for sources of the same sign (for instance, two identical branes), and attractive for sources of opposite sign (for instance, a brane–antibrane system), just like the interaction between two electric charges. In particular, for the system formed by two identical, static and parallel branes (like those appearing in the initial configuration of the ekpyrotic scenario), one finds that the gravitational attraction is exactly balanced by the repulsion due to the antisymmetric-field interaction, and the system remains static (up to the addition of other, nonperturbative interactions).

An inflationary model can be obtained, in the context of cosmological models based on Dirichlet branes if we consider the interaction of a D_3-brane with an anti-D_3-brane. In that case, there is no cancellation between the various types of force, and the net result is an attractive interaction between the branes. The coordinate parametrizing the interbrane distance behaves as a scalar field, and its potential (generated by the forces between the branes) can in principle sustain a phase of inflationary expansion, as pointed out by various groups of theoretical physicists and astrophysicists (including Gia Dvali, Henry Tye, Clifford Burgess, Mahbub Majumdar, Detlef Nolte, Fernando Quevedo, Govindan Rajesh, and Ren-Jie Zhang).

Unfortunately, if the external dimensions orthogonal to the branes are flat and topologically trivial, it turns out that the effective inflationary potential generated by the brane–antibrane interaction is unable to guarantee a successful resolution of all the standard cosmological problems. However, there are two ways out of this difficulty, at least in principle.

A first possibility relies on the assumption that the space transverse to the branes is compact and has the topology of an n-dimensional torus, with spatial sections of uniform radius r. When the separation of the brane–antibrane pair is of order r, the effective potential experienced, say, by the antibrane must be estimated by including the contribution of all the topological "images" of the other brane, forming an n-dimensional lattice. The total effective interaction is then obtained by summing over all the contributions of the lattice sites occupied by the brane images. The resulting effective potential can satisfy the conditions for successful inflation (at least, as long as the interbrane separation remains in a range of distances of order r).

An alternative solution (which seems to be preferred, at present, in view of its ability to stabilize the size of all the additional compact dimensions present in the model) has been suggested by Shamit Kachru, Renata Kallosh, Andrei Linde, Juan Martin Maldacena, Liam McAllister, and Sandip Trivedi. Their model is based on the assumption that the section of space orthogonal to the D_3-branes is curved, with a geometry of anti-de Sitter type (as in the case of the Randall–Sundrum model). One of the two branes (for instance, the antibrane $\overline{D}_3$) is frozen at a fixed position. The D_3-brane, on the other hand, is mobile along the orthogonal direction z, driven by the attractive force towards the antibrane, and has a time-dependent position (see Fig. 10.4). The potential generated by the interbrane interaction is a function of the (time-dependent) interbrane distance, exactly as in the the case of a flat geometry. However, the potential energy is now "distorted" by the curvature, which produces a warping of the spatial geometry along the z direction. As a consequence, the effective potential acquires a form satisfying all conditions required to implement a successful inflationary model.

Let us conclude this chapter by noting that, in all models of brane–antibrane inflation, as in the case of the ekpyrotic scenario, the initial configuration settles down in the strong-coupling, M-theory regime, within an eleven-dimensional space-time filled with strings, membranes, three-branes, and so on, including the whole possible "zoo" of higher-dimensional objects in mutual in-

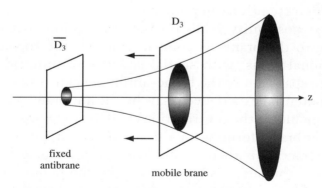

FIGURE 10.4 Schematic view of the interaction between a brane D_3 and an antibrane $\overline{D}_3$ embedded in a curved anti-de Sitter geometry. The brane is attracted towards the antibrane, located at a fixed position. The funnel aligned along the z direction illustrates the shrinking of the so-called warp factor (i.e., the effective gravitational redshift) produced by the external curvature upon the four-dimensional geometry of the brane

teraction. As the value of the initial coupling decreases, however, those heavy higher-dimensional objects tend to disappear from the initial state. In the limiting case in which the initial coupling tends to zero, we recover the string perturbative vacuum, a state without strings or branes, where there is nothing at all in the flat, cold, and infinite space-time but the unavoidable microscopic quantum fluctuations of the metric and the other background fields (the initial state described in Chap. 5).

The specific model of cosmological evolution thus crucially depends on our assumptions about the initial value of the coupling parameters (i.e., about the initial strength of all interactions). Different initial conditions can lead to different types of cosmological evolution, and different ways to reach the present cosmological state. Large-scale astrophysical observations, which are becoming more and more accurate, will soon be able to reconstruct the past history of our Universe, thus providing indirect information about the strings (or membranes, or three-branes, etc.) that were present at the beginning.

11. Conclusion

The key message we hope to have transmitted to the reader in this book is that there are neither observational data nor incontrovertible theoretical arguments to support the belief that the Big Bang represents the beginning of the Universe, and that before the explosion there was "nothing". On the contrary, there are solid motivations – based upon recent developments in theoretical physics – for thinking otherwise. There are also scientifically valid tools for tracing the history of the cosmos back to epochs preceding the Big Bang, providing ways of testing such investigations with a series of effective experimental observations that are already feasible, and with even better prospects in the near future.

The kinematical and dynamical details of the primordial cosmological epochs preceding the Big Bang are still quite uncertain. There are various models, many hypotheses, and a number of possibilities that have not yet been fully explored. The situation will certainly become clearer following the theoretical and experimental work to be carried out over the next few decades. It is already evident, however, that the Big Bang could lose its rather mystical role as the beginning of *everything*, to become a more modest beginning of the *current phase of the Universe*, i.e., of the Universe as we currently know it, made up of radiation, matter, atoms, galaxies, and human beings. Nevertheless, it would still represent a crucial step in the history of our Universe, without which life in the form we now experience it would probably be absent.

With regard to the beginning of the Universe, it is amusing to consult the book that can probably be considered the first and most authoritative text on cosmology of the modern age: *Genesis*. Actually, by carefully rereading the initial verses of the Holy Bible, we find a description of the birth and the first moments of our Universe which seems much closer to the pre-Big-Bang scenario than to the standard Big Bang scenario. Indeed, there is no mention at all of an explosion, and no reference to any hot, dense, highly curved concentration of energy. What is described is rather an initial state that is completely quiet, deserted, dark, and lifeless, which just resembles the typical initial state of pre-Big-Bang models (described in poetic, but very appropriate, terms). We can read, in fact:

> First God made heaven and earth.
> The earth was without form and void,
> and darkness was upon the face of the deep;
> and the Breath of God
> was moving over the face of the waters...
>
> (Genesis, The Holy Bible)

Here, terms like "heaven" and "earth" could denote, respectively, space-time itself – i.e., the environment where the Universe is brought to life and subsequently evolves – and the various forms of energy and natural forces. The "darkness" and the "deep" give us the idea of something immensely large, empty, and cold, like empty space, void of any interaction. Indeed, without interactions, matter is dark, since it does not emit radiation (i.e., light). The whole scenario actually makes us think of the string perturbative vacuum, which is a free state (i.e., without interaction) with a flat space-time geometry.

We should also recall that the fundamental string coupling determining the strength of all forces is controlled by the dilaton, and in particular by the exponential function of the dilaton field (see Chap. 4). In order to have an arbitrarily small coupling (i.e., arbitrarily weak interactions) in the initial state, the initial value of the dilaton field must be arbitrarily large and negative: this huge negative "abyss" could correspond – with a little imagination – to the "deep" mentioned in the Genesis.

Furthermore, "without form" is approximately synonymous with incoherent, chaotic, or stochastic. But this fits quite well with the description of the initial conditions given at the end of Chap. 5, where we presented an analogy between such an initial state and an ocean whose waves collide chaotically, and occasionally trigger non-trivial physical processes. The breath of life over this "face of the waters" could just represent the quantum oscillations of the dilaton and the geometry, aimed at triggering the inflation mechanism that eventually brings the Universe to its standard configuration through the explosive stage of the Big Bang.

Of course, everybody knows that the Bible's words cannot be taken literally. It is also well known that it would be misleading to give a subjective interpretation to those words, forcing their meaning to fit one's opinion. Nevertheless, it is difficult to refrain from proposing a personal translation of the above verses in scientific terms, those verses which so poetically describe the origin of the Universe in a language appropriate to ancient times when Genesis was written. Using a modern, less metaphorical language, the translation could sound more or less like this:

> First God made the fields and the sources.
> The sources were incoherent in the vacuum,
> and this dark matter was without interactions;
> and the dilaton
> was fluctuating over the string perturbative vacuum ...

The next sentence: And God said: let there be light! seems to describe the Big Bang, i.e., the production of radiation marking the beginning of the standard cosmological phase! Hence, according to this personal translation, Genesis describes a scenario for the creation that seems to correspond quite closely to the pre-Big-Bang scenario suggested by string cosmology. But, as everybody knows, one can read anything into the Bible, provided one looks carefully enough for it.

More seriously, it would be naive to ask from science an explanation for all the big question marks of the creation. Beyond a certain point, each of us should look into himself/herself for the answers to the fundamental questions pertaining to the existence of the Universe and our own existence. My personal and modest opinion (as far as it may count in this case) is that the Universe was born according to God's will, with an act of creation having its ultimate and complete purpose in human beings. However, with regard to how the Universe evolves after its creation, following those laws that God himself wanted to instill into nature, I think it is fully appropriate to apply the methods of scientific investigation.

In this spirit, string theory applied to cosmology seems to tell us that the Big Bang is *not* to be identified with the time of the initial creation, in the same way that – if I can take the liberty of using an analogy from biology – childbirth must not be identified with the moment when a new life is created (which corresponds rather to the act of conception). Well before the Big Bang, the supernatural act of creation was followed by a long cosmological "pregnancy", required to prepare the explosion leading the Universe to its current form (similarly to what happens after the conception of a new living creature during the time preceding delivery). Actually, we may think of the Universe before the Big Bang as being in a sort of embryonic state, during which the various physical properties (that will be made manifest later on, during the post-Big-Bang epochs) were gradually taking form.

This "prenatal" life of the Universe is fully accessible to present and future experimental investigation. The hypothesis of a self-dual Universe, the scenarios described by string cosmology and brane cosmology models, and so on, can be tested in various ways. We may recall, in particular, that the phase of pre-Big-Bang evolution may produce backgrounds of relic gravitational radiation much stronger than the ones predicted by standard cosmology at high frequencies, and hence more accessible to direct observation. In addition, the typical production of electromagnetic seeds for the cosmic magnetic fields, and of axion seeds for the CMB anisotropy, could very soon lead to other possible (even if indirect) confirmations or disproofs. Current and near-future experiments are thus able to open a window on the earliest history of the Universe, on epochs much more remote than ever envisaged. This is the other important message that this book hopefully puts across.

It seems appropriate to conclude with a historical remark. The current status of cosmology, characterized by various possible models for the primordial Universe, looks similar to the situation about half a century ago, when there were two contrasting cosmological scenarios. They were somehow complementary, and corresponded to two radically different visions of the cosmos: the steady-state Universe of Herman Bondi, Thomas Gold, Fred Hoyle, and Jayant Narlikar, characterized by a continuous creation of matter, and the evolutionary Universe, hot and explosive, born from the Big Bang, of Georges Lemaitre, George Gamow, Robert Dicke, and others. One of the crucial differences between these scenarios was, respectively, the absence and the presence of a cosmic background of thermal radiation. It was just the direct observation of this background, discovered by Arno Penzias and Robert Wilson in 1965, that definitively confirmed one scenario and disproved the other.

The current situation is rather similar. There are standard inflationary models predicting a very low, practically undetectable background of cosmic gravitational radiation at high frequencies. Other models, based upon string theory, predict a much higher background. Once again we expect the choice between the different scenarios to be made on the basis of experiment, hopefully in a not too distant future.

Whatever the answer, we do believe that the experimental study of the relic gravitational background will be as important for cosmology as the study of the electromagnetic microwave background. Probably even more important, since the electromagnetic radiation contains photons which provide us with a snapshot of a Universe younger than the current one, but still subsequent to the Big Bang. The relic gravitons, on the other hand, originate from a much more remote past, and may have retained in their spectrum a permanent imprint of the pre-Big-Bang Universe.

References

An excellent introduction to the physics of the standard cosmological model for non-specialized readers can be found in the following book, written by the Nobel Prize laureate Steven Weinberg:

1. S. Weinberg: *The First Three Minutes* (Basic Books, New York 1977)

For a more technical overview of the same topics, completed by a pedagogical exposition of general relativity, a good reference is the following textbook by the same author:

2. S. Weinberg: *Gravitation and Cosmology* (Wiley, New York 1972)

A vast and accessible introduction to standard inflationary cosmology, also presented with a historical perspective, can be found in the popular book written by Alan Guth, considered to be one of the fathers of the inflationary scenario:

3. A. Guth: *The Inflationary Universe* (Vintage, London 1997)

For a more technical and quantitative discussion of the transition from standard to inflationary cosmology, and the basic aspects of the inflationary scenario, there is the excellent textbook written by two well known astrophysicists at the University of Chicago:

4. E.W. Kolb and M.S. Turner: *The Early Universe* (Addison-Wesley, Redwood City, Ca 1990)

For inflation we also refer the reader to the more recent book written by two other well-known English astrophysicists, Andrew Liddle and David Lyth, which gives a professional illustration of various inflationary models and their observational consequences:

5. A.R. Liddle and D.H Lyth: *Cosmological Inflation and Large-Scale Structure* (Cambridge University Press, Cambridge 2000)

Readers particularly interested in the physics of CMB radiation are also referred to the book written by the astrophysicsts Amedeo Balbi (an expert in the field of the CMB radiation), and easily accessible to the non-specialized reader:

6. A. Balbi: *The Music of the Big Bang* (Springer, 2008)

For a first, qualitative approach to gravity, black holes, and quantum cosmology, there are two popular books by Stephen Hawking, probably one of the most famous theoretical physicists in the field of gravity and cosmology:

7. S.W. Hawking: *A Brief History of Time* (Bantam Books, 1988)

8. S.W. Hawking: *Black Holes and Baby Universes* (Bantam Books, 1993)

For string theory, it is fair to say that studies are still in progress. However, an excellent overview for non-specialized readers of the current status of string theory and of its main physical applications can be found in the book by Brian Greene:

9. B. Greene: *The Elegant Universe* (Vintage, London 1999)

A fully detailed and professional introduction to string theory can be found in the following two books, written by some of the most famous experts in this field:

10. M.B. Green, J. Schwartz, and E. Witten: *Superstring Theory* (Cambridge University Press, Cambridge, 1987)

11. J. Polchinski: *String Theory* (Cambridge University Press, Cambridge, 1998)

For more details about string cosmology, and a more technical presentation of the pre-Big-Bang scenario, the interested reader can refer to the professional review paper written by Gabriele Veneziano (one of the fathers of string theory) and the present author:

12. M. Gasperini and G. Veneziano: *The Pre-Big Bang Scenario in String Cosmology*, Phys. Rep. **373**, 1 (2003)

There is also a recent, self-contained textbook written by the present author, offering a pedagogical introduction to string cosmology, primarily intended for graduate students of theoretical physics and astrophysics (but not too specialized). It includes the pre-Big-Bang scenario and all the other ideas discussed in this book, and summarizes the progress obtained in this field over the last fifteen years:

13. M. Gasperini: *Elements of String Cosmology* (Cambridge University Press, Cambridge 2007)

For those surfing the net we also suggest the web page devoted to string cosmology, available at the address below. This site has existed since 1995 and is regularly updated. It is an archive of papers published in scientific journals and mainly concerned with pre-Big-Bang cosmology. The papers are classified by subject, and the site contains the following main sections (each of them divided into various subsections): (1) The general picture. (2) Motivations. (3) Fundamental problems. (4) Phenomenological aspects. (5) Aspects of quantum cosmology. (6) Generalizations. (7) Review talks and lectures. The papers included in these sections are either available for direct download, or linked to the arXiv e-print service SPIRES-HEP at SLAC (Stanford):

14. http://www.ba.infn.it/~gasperin

Finally, we refer readers to the useful Particle Data Group web page at the following address. This provides a very accurate database containing the numerical values of all parameters of particle physics, cosmology, and astrophysics, to the best accuracy available by current measurements. All observable physical quantities often used (or mentioned) in this book (like the speed of light c, the present Hubble parameter H_0, the Newtonian constant G, the present temperature T_0 of the CMB radiation, the masses of the various particles, etc.) can be found within the Particle Data Group catalogue, together with their precise definitions and up-to-date numerical values:

15. http://pdg.lbl.gov

Index